ENVIRONMENTAL SCIENCE, ENGINEERING AND TECHNOLOGY SERIES

POTENTIAL OF ACTIVATED SLUDGE UTILIZATION

ENVIRONMENTAL SCIENCE, ENGINEERING AND TECHNOLOGY SERIES

Nitrous Oxide Emissions Research Progress
*Adam I. Sheldon
and Edward P. Barnhart (Editors)*
2009. ISBN: 978-1-60692-267-5

Fundamentals and Applications of Biosorption Isotherms, Kinetics and Thermodynamics
Yu Liu and Jianlong Wang (Editors)
2009. ISBN: 978-1-60741-169-7

Environmental Effects of Off-Highway Vehicles
*Douglas S. Ouren,
Christopher Haas, Cynthia P. Melcher, Susan C. Stewart, Phadrea D. Ponds,
Natalie R. Sexton Lucy Burris, Tammy Fancher and Zachary H. Bowen*
2009. ISBN: 978-1-60692-936-0

Agricultural Runoff, Coastal Engineering and Flooding
*Christopher A. Hudspeth
and Timothy E. Reeve (Editors)*
2009. ISBN: 978-1-60741-097-3

Agricultural Runoff, Coastal Engineering and Flooding
Christopher A. Hudspeth and Timothy E. Reeve (Editors)
2009. ISBN: 978-1-60876-608-6
(Online book)

Conservation of Natural Resources
Nikolas J. Kudrow (Editor)
2009. ISBN: 978-1-60741-178-9

Conservation of Natural Resources
Nikolas J. Kudrow (Editor)
2009. ISBN: 978-1-60876-642-6
(Online book)

Directory of Conservation Funding Sources for Developing Countries: Conservation Biology, Education and Training, Fellowships and Scholarships
Alfred O. Owino and Joseph O. Oyugi
2009. ISBN: 978-1-60741-367-7

Forest Canopies: Forest Production, Ecosystem Health and Climate Conditions
Jason D. Creighton and Paul J. Roney(Editors)
2009. ISBN: 978-1-60741-457-5

Soil Fertility
Derek P. Lucero and Joseph E. Boggs (Editors)
2009. ISBN: 978-1-60741-466-7

The Amazon Gold Rush and Environmental Mercury Contamination
Daniel Marcos Bonotto and Ene Glória da Silveira
2009. ISBN: 978-1-60741-609-8

Process Engineering in Plant-Based Products
Hongzhang Chen
2009. ISBN: 978-1-60741-962-4

Buildings and the Environment
Jonas Nemecek and Patrik Schulz (Editors)
2009. ISBN: 978-1-60876-128-9

Tree Growth: Influences, Layers and Types
Wesley P. Karam (Editor)
2009. ISBN: 978-1-60741-784-2

Potential of Activated Sludge Utilization
Xiaoyi Yang
2010. ISBN: 978-1-60876-019-0

ENVIRONMENTAL SCIENCE, ENGINEERING AND TECHNOLOGY SERIES

POTENTIAL OF ACTIVATED SLUDGE UTILIZATION

XIAOYI YANG

Nova Science Publishers, Inc.
New York

Copyright © 2010 by Nova Science Publishers, Inc.

All rights reserved. No part of this book may be reproduced, stored in a retrieval system or transmitted in any form or by any means: electronic, electrostatic, magnetic, tape, mechanical photocopying, recording or otherwise without the written permission of the Publisher.

For permission to use material from this book please contact us:
Telephone 631-231-7269; Fax 631-231-8175
Web Site: http://www.novapublishers.com

NOTICE TO THE READER

The Publisher has taken reasonable care in the preparation of this book, but makes no expressed or implied warranty of any kind and assumes no responsibility for any errors or omissions. No liability is assumed for incidental or consequential damages in connection with or arising out of information contained in this book. The Publisher shall not be liable for any special, consequential, or exemplary damages resulting, in whole or in part, from the readers' use of, or reliance upon, this material.

Independent verification should be sought for any data, advice or recommendations contained in this book. In addition, no responsibility is assumed by the publisher for any injury and/or damage to persons or property arising from any methods, products, instructions, ideas or otherwise contained in this publication.

This publication is designed to provide accurate and authoritative information with regard to the subject matter covered herein. It is sold with the clear understanding that the Publisher is not engaged in rendering legal or any other professional services. If legal or any other expert assistance is required, the services of a competent person should be sought. FROM A DECLARATION OF PARTICIPANTS JOINTLY ADOPTED BY A COMMITTEE OF THE AMERICAN BAR ASSOCIATION AND A COMMITTEE OF PUBLISHERS.

LIBRARY OF CONGRESS CATALOGING-IN-PUBLICATION DATA

Yang, Xiaoyi.
 Potential of activated sludge utilization / Xiaoyi Yang.
   p. cm.
 Includes bibliographical references and index.
 ISBN 978-1-60876-019-0 (hardcover)
 1. Digester gas. 2. Gas as fuel. 3. Sewage sludge fuel. 4. Sewage--Purification--Anaerobic treatment. I. Title.
 TD769.4.Y36 2009
 628.3'54--dc22
                                                            2009027060

Published by Nova Science Publishers, Inc. ✣ *New York*

CONTENTS

Preface		ix
Chapter 1	Introduction	1
Chapter 2	Potential of Anaerobic Sludge Digestion	9
Chapter 3	Potential of Sludge Pyrolysis	31
Index		47

PREFACE

As the lack of fossil fuel is increasing, sludge management is not only to satisfy the disposal criteria but also to obtain energy and resource. For a better sewage sludge disposal and more efficient energy reclamation, a series of analysis and experiments were performed to study the potential of anaerobic digestion.

The biogas outputs in waste sludge treatment. Anaerobic digestion techniques have traditionally been employed to reduce the volume and weight of sludge and produce corresponding amounts of biogas. As most of the organics present in sewage sludge are enveloped by low biodegradability of the cell walls and extracellular biopolymers, the rate-limiting step in sludge digestion is generally believed to be the hydrolysis of particulate organic matter to soluble substance. Novel pre-treatment processes, WAO can improve biogas production. The two-stage first-order reaction kinetic model and the generalized kinetic model are applied to study the kinetics of the wet air oxidation process of industrial waste activated sludge, and the rate of MLVSS/MLSS was used as the model parameter instead of COD or TOC to shorten the error that results from sampling. The generalized kinetic model is relatively available to predict the WAO process of waste sludge. The values of point-selectivity show that there is a strong presence of acetic acid and short chain organic substance in the WAO treatment process, which were experimentally confirmed by chromatogram-mass spectrograph instrument.

Chapter 1

1. INTRODUCTION

Activated sludge processes are the main alternative to wastewater treatment due to their low cost and high efficiency. However, activated sludge processes may lead to a solid waste disposal problem because of the production of sewage sludge, which increases with increasing amounts of wastewater being treated. In turn this leads to increasing difficulties in managing sludge as a resource for energy production and meeting increasingly strict criteria for disposal. Therefore, it is a very important and complex problem to manage waste activated sludge (WAS) in an economically and environmentally acceptable manner. The disposal requirement and present sludge disposal are introduced in this chapter and existing problems in sludge management are also surveyed.

1.1. DISPOSAL REQUIREMENTS

Sludge disposal is much complicated because of the complex compositions in sludge. At present, sludge disposal techniques include land application, landfill, disposal into the sea and incineration.

The requirements for sludge disposal have gradually become more rigorous. Disposal into the sea was strictly prohibited in China. At beginning of 1994, Chinese government accepted three international agreements and committed not to discharge industrial wastewater and waste sludge into the sea (Qiao, 2000). In Europe, sludge into the sea is currently practiced only by three countries, Ireland 35%, UK 30% and Spain 10%, which is still controlled in accordance with certain requirements (e.g. the Oslo Convention for the protection of the North Sea and the North East Atlantic) and even licences issued under national legislation which take into account the quantity and quality of the sludge and the nature of the receiving area (Mathews, 1992). In Europe, there will be no more dumping of sludge into the ocean as from 31 December 1998, when the North Sea Conference Agreement comes into force and sea disposal is banned(Goldsmith,1991). In USA, ocean disposal by states having coastal line has been practised, but this was expected to be phased out because of changes in water pollution control regulation(McGhee, 1991).

Land application was ever considered the most promising management. However, as the consequences of irrational sludge utilization, many countries had established correspondingly laws and rules to minimize its harm and make a rational use. In the United States, standards for sludge land application list the highest concentrations of heavy metals, pathogens and organic substances. Moreover, the total highest concentrations of contaminants accumulated in soil and contaminant concentrations per unit land were limited in the requirement. The European Union also recommended the standards for nutrient concentrations, heavy metal contents and even sludge dosage per year.

In 1984, in terms of the studies on pollutant control standard of sludge applied for agriculture land, China came out sludge agriculture land standard (GB-4284-84), which had effects on controlling irrational application of sludge on agriculture land. Although the standards of different countries in sludge applied for agriculture use are somewhat different and various nations pay attention to different heavy metal in sludge, sludge for agriculture use faces growing difficulties due to increasing stricter disposal standards.

1.2. RECENT SLUDGE DISPOSAL

Sludge disposal options available should be in accordance with the local geographic condition, the local climate, the local economic level and transportation besides of the law and the rules. Every kind of management in the actual applications has some advantages with the weakness in practical application.

Disposal into the sea operates simply and cost lower especially for the coast city, but it may bring some unpredictable problems, so that it was strictly prohibited in many counties.

Incineration takes 24% of the sludge produced in Denmark, 20% in France, 15% in Belgium and 14% in Germany. In the USA and Japan, 25% and 55% of the sludge produced, respectively, is incinerated

Disposal of sewage sludge to sanitary landfills still takes the bulk of sludge in developed countries. About 40% of the sludge produced in the European Union is disposed of through land filling. For Greece, Luxembourg and Italy, where 90, 88 and 85%, respectively, of the sludge is taken to landfills. In the U.S.A., 48% of the sludge produced in 1990 was deposited through land filling (McGhee, 1991). However, it is becoming to be limited due to its highly costs to establishing sanitary landfill sand difficulties for finding appropriate sites.

A valid treatment method should make the balance between the environment benefits and economic costs. In European Union and America, sludge disposal still gave priority to agriculture use, and the rate of sludge for agriculture use increased in European Union. From the point of economic factor and resource reuse, as a developing country, sludge application for agriculture use is in accordance with the native situation. In fact, municipal sludge in China has regarded agriculture use as principal method since 60's, 90% of municipal sludge was used for agriculture land. In 1990, there was only 78 municipal wastewater treatment plants in China, and in only 25% of plants, the sludge had been stabilized. In 1993 among 119 municipal wastewater treatment plants, 55 municipal wastewater treatment plants (occupying 46.2%) had treated sludge (Rulin, 1994). Although the quality of digestion sludge is improved, the problems about its nutrient concentration lowness and heavy metals contamination still exist, and sludge for agriculture use still face difficulties.

Efficiencies of sludge applied for land among different regions were quite obviously different due to characteristic of sludge and soil. The sludge from domestic wastewater treatment plants in Guilin, China, contained a relative high organic matter, nitrogen, phosphorus and potassium. Moreover, the heavy metals satisfied the national standard of sludge for agricultural land. Therefore, sludge was appropriate for land application and even was mixed with additive and packed as an organic complex fertilizer. The results of organic complex fertilizer for agriculture use indicated that the rice yield was increased by 19% and was a little higher than or equal to the normal complex fertilizer. The output of sugarcane increased 22% in contrast with normal complex fertilizer and increased 29% compared with the complex fertilizers containing calcium magnesium phosphate, carbamide and potassium chloride.

Complex granule fertilizers, which are produced by sludge, chemical substances and additive, are an option for sludge management, which not only compensate costs on sludge management but also enhanceing nutrient contents for sludge application use.

In southeast of Beijing, there were many years on sludge applied for agriculture use. In 1977 and in 1992, it was respectively investigated that part of farmlands and farm crops already suffered pollution of heavy metals in different levels. Moreover, the more times municipal sludge was applied in the land, the more serious rice and vegetable were polluted. Cd was in excess of national hygienic standard.

At present, the problems of disposal and utilization of sludge are necessary to be solved: 1.How to solve the second pollution induced by sludge disposal like heavy metals, pathogens in sludge? 2. How to get more energy from sludge treatment? 3. How to make appropriate use of sludge and satisfy the laws and rules at the same time? The studies are being carried out for solving those questions.

1.3. PROGRESS IN THE RESEARCH OF MUNICIPAL WASTE SLUDGE MANAGEMENT

1.3.1. Rational Application and Science Strategy

In 60's, municipal sludge from Beijing Gaobeidian wastewater treatment plant was applied for land near the plant. Afterward tens years, the technologies of sludge management developed rapidly, from 1985 to 2000, many research institutes and universities have made many progresses of sludge management including sludge rational application for land, sludge complex fertilizers, sludge potential harm and prevention, and cure strategy.

Perfecting and Executing Standard of Pollutants Control
The standard of sludge for agriculture use in China has more strict limits on heavy metals in sludge, and is referred available heavy metals in sour or alkaline soil, but it lacks the standard of time and dosage control. Only to control the highest concentrations of heavy metals is not quite enough as the results of following reasons: heavy metals are inclined to accumulate in soil, and availability of different heavy metal is quite different to different pH of soil, moreover, it is affected by other conditions including organic concentration, cation exchange capacity, characteristic of soil, mineralization contents of soil.

Qiao Xianliang(2000) presented that the concentrations of heavy metals in soil and in sludge co-controlled the application of sludge. Dosage of sludge for land every year should depend on plant need for nutrient and nutrient total contents in sludge. He pointed that once dosage was used more than need, nitrogen in sludge was inclined to leak out and may cause water eutrophication and nitrite contamination, moreover he considered phosphorus could be used as a second consideration factor due to less pollution for its weak ability to move.

The research project of " Wastewater and Sludge Utilization Technique for Agriculture" was made by Beijing environment science research institute in 1990~1995(Ke Jianmin et al.,2000). The amount of sludge applied for land depends on input of restricted soluble nutrient and heavy metals. In order to fully utilize nutrient in sludge and at the same time avoid nutrient and heavy metals contamination. The concept of sludge application efficiency was put forward, including three kind of concept of sludge application efficiency, namely, once the highest applied quantity, safely applied quantity and control-applied quantity.

From the report of Wang Hongkang(1990), heavy metals accumulation made products of rice, wheat and Soya bean decreased due to sludge for land, and he suggested that control standard of sludge applied for land should be Pb 60mg/kg, Ni 330mg/kg, and on alkaline soil Cu 800mg/kg. Cao Renlin (1997) showed that nitrate movement from the surface to the underground was the major pollution factor, he pointed it would not cause adverse effects on surface water and underground water when the dosage of sludge applied for land was controlled within 50 tons hm^{-1}

Removal Heavy Metals

The presence of heavy metals such as Zn, Cu, Cr, Cd, Pb beyond recommended standard is one of the important obstacles to its general use. Therefore, chemical and bioleaching methods were used to remove heavy metals from municipal sludge.

Chemical method can get 80~90% metal removal efficiency, but it is difficult to use in practice in China because of high cost, operational difficulties and large consumption of chemical agents. However, sometimes only adjustment of pH of sludge, most of heavy metals can loss their availability. Wen yanmao (1999) adjusted sludge by CaO, Ni and Zn available contents in soil decreased obviously while Cr increased, and Ni, Zn, Cd in the plants decreased in large scale. The results showed that adjusting the pH of sludge to 6.5~7.5 can control availabilities of Ni, Zn and Cd in soil and absorption of Ni, Zn and Cd in plant.

Bioleathing (Zhou Shungui et al., 2002) appears to be more practical and economical for metals removal. Under optimum bioleaching conditions, Zn, Cd, Ni and Cu can be almost completely removed, whereas removal efficiencies of Pb and Cr are lower. Metal sulfides can be oxidized into soluble sulfates by bacteria, and then removed by leaching. Bioleaching has several advantages compared to chemical methods, such as reducing acid consumption by more than 80%, dewatering easy, reducing fertilizer value loss in municipal sludge. Zhou lixiang

(2000) divided municipal sludge into 4 components (biofloculent, particulate, colloid and soluble) by precipitation centrifuge method on different rotational speed and studied the distribution of heavy metal in sludge. It was found that in particulate component almost all metals were contributed to the less available forms. It is concluded that available amount of sludge-born metals in sludge-treated soil depended on, to great extent, the decomposition of the sludge biofloc component. In batch experiments (Zhou Lixiang and Wang Genmin, 2001), 93% of Cu and 85% of Zn could be removed from sludge by bioleaching.

1.3.2. Reducing Sludge Output

Waste sludge management would be at a correct balance between inhibitory and disposal sludge. This balance should give priority to the output of waste sludge reducing. Liu(1998) interpreted metabolic behaviors of activated sludge microorganisms under substrate-sufficient condition, and established a series of bioenergetics models describing the relationship between microbial anabolism-catabolism and S_0/X_0 ratio, and these models satisfied experimental data. Liu (1998) also showed a novel technique for solving excessive sludge management, which was able to greatly reduce sludge production. Experiment results showed that more than 60%~80% of substrate was consumed through energy uncoupling, namely, the sludge product would decrease by 60%~80% according to this theory.

Sludge disposal of wastewater treatment plant is an important subject of environmental protection in China. The production of municipal sludge is increasing rapidly along with economic development. Land application is extensively applied for treating municipal sludge. Heavy metal is the key factor influencing the sludge for agriculture use in China. Many researches had been made to develop the technologies of reducing sludge output, reuse and low cost and high efficiency management.

REFERENCES

Cao Renlin et al. 1997. Research on effect of sewage sludge compost applied for green garden. *Environment Science Research* 10(3), 46-50.

Cecil Lue-Hing. 1996. Peter Matthews et al. Sludge management in highly urbanized aeras. *Wat. Sci. Tech.* 34(3-4), 517-524.

Davis.R.D. 1996. The impact of EU and UK environmental pressuers on the future of sludge treatment and dispasal. *Water Environ. Manage* 10(2), 65-69.

Ding Wenchuan, Hao Yiqiong, Tang Zihua. 2000. Sludge treatment and disposal of Chongqing municipal wastewater treatment plants. *Chongqing Environmental Science* 22(4), 14-17

Goldsmith P. 1994. Sanitary solutions. *The Chemical Engineer* 13–14.

Jin Rulin, 1994. Discussion of sludge treatment of urban wastewater plant in china. *J.Wuhan Urban Construction Institute* 11(2), 1-12.

Ke Jianming, Wang Kaijun, Tian Ningning. 2000. Treatment and disposal of excess sludge of Beijing urban wastewater treatment plant. *China Biogas* 18(3), 35-38.

Li Guibao, Yin Chengqing. 2001. A view of the progess of sewage sludge use in forestland and green aeras in China. *J. of Beijing Forestry University* 23(4), 71-74.

Liu Yu, Wang Qidong. 1998. Novel technique of controlling excessive sludge production in activated sludge process. *J. Beijing institute of Light Industry* 15(3), 39-43

McGhee T.J. 1991. Water supply and sewerage. New York: McGraw-Hill.Pang Jinhua. 1994. Effect of sewage sludge on regional ecoenvironment. *Tropical and Subtropical Soil Science* 3(1), 41-47

Qiao X., Luo Y., Wu S., 2000. Application in agriculture and its affects of sewage sludge. *Soil* 2, 79-85.

Qiao X., Luo Y., 2001. The chemical content and ordinance agriculture of urban sludge in China. *Soil* 4, 205-224

Tian N., Wang K., 2000. Technology Evaluation on disposal and utilization of Excess Activated sludge. *Engineering and Technology* 2, 18-20

Wang D., Xie Q., 1999. Study on agricultural utilization of the sludge disposal of municipal wastewater. Chongqing Enviro. *Science* 21(6), 50-52.

Wang D., Xie Q. 1999. A preliminary study on sludge disposal of municipal waste water. *Journal of Guilin Institute of Technology* 19(4), 387-390.

Wang H., 2000. Investigation and recent status on manure preparation wastewater sludge. *Water and Wastewater* 26(9), 1-3

Wang H.,Yan S., 1990. Research of crop pollution of Cu for applied sewage sludge. *Enviroment Science* 11(3), 6-11.

Wang S., Cui Y., 2000. Development of the sludge treatment techniques. *J. Jilin Construction Engineering Institute* 3, 25-28.

Wei C., Chen C., 1998. The analysis of current research on the treatment disposal and utilization of sludge. *Urban Environment and Urban Ecology* 11(4),10-13

Wen Yanmao, Liu Yanbin. 1999. Study on availability control of heavy metals in soil applied with municipal sewage sludge. *J. Zhongshan University* 38(4), 97-101

Wen Yanmao, Wei Zhaotao. 1996. Concentration and availability of heavy metals in municipal sewage sludge and soil in Guangzhou. *Journal of Zhongshan university* 35, 217-221

Xie Qinglin, Wang Dunqiu. 2000. Agricultural Utilization of the sludge of sewage treatment plants in China. *Journal of Guangxi academy of science* 16(3),131-134.

Yang Zhouya, Wang Hongkang. 1993. Research of crop pollution of Pb for applied sewage sludge. *Enviroment Science* 14(6), 8-11,37

Yao Gang. 2000. Sewage sludge treatment, Use and Disposal Germany. *Urban Enviroment and Urban Ecology* 13(1), 43-47

Zhang Qiao, Wu Qitang. 2000. Effect of sludge compost application on crop and soil. *Soil and Enviroment Sciences* 94(4), 277-280

Zhang Qingmin, Chen Weiping. 2000. State and development and Disposal of sewage sludge in city. *Agro-environmental Protection* 19(1), 58-61.

Zhang Suxia, Wang Hongkang. 1991. Research of crop pollution of Ni for applied sewage sludge. *J. of Enviroment Science* 11(1), 71-78.

Zhao Lijun, Zhang Daqun, Cheng Baozhu. 2001. Development in sludge treatment and disposal. *China Water and Wastewater* 17(6), 23-25.

Zhou Lixiang, Hu Aitang, Ge naifen. 1994. Composition of municipal sewage sludge and effects on vegetables and soil. *Journal of Nanjing Agricultural University* 17(2), 54-59.

Zhou Lixiang, Shun Qirong. 2000. Distribution and chemical form of heavy metals in the principal component of undigested sludge. *Enviromental Science Transaction* 20(3), 269-274.

Zhou Lixiang, Wang Genmin. 2001. Bioleaching of heavy metals from sewage sludge. *Enviromental Science Transaction* 21(4), 504-506.

Zhou Shungui, Zhou Lixiang. 2002. Removal of heavy metals from sludge by bioleaching. *ACTA Ecological Sciencea* 22(1), 125-133.

Zhou Yimin, Zhang Jinsheng. 1990. Investigation on concentration of heavy metals in vegetable and soil using municipal sewage sludge in Tianjin. *Agriculture Environmental Protection* 9(6), 30-34

Chapter 2

2. POTENTIAL OF ANAEROBIC SLUDGE DIGESTION

Although different possibilities exist for waste sludge treatment, anaerobic digestion plays an important role for its ability to produce energy-rich biogas, to destroy pathogens and to stabilize the sludge (Lise, 2008). Anaerobic digestion techniques have traditionally been used to reduce the volume and weight of sludge and to produce biogas in spite of the process being limited by long retention times and the low overall degradation efficiency of the organic matter. As most of the organics in sewage sludge are involved in poor biodegradable cell walls and extracellular biopolymers, the rate-limiting step in sludge digestion is generally believed to be the hydrolysis of organic matter. Thus various pretreatment processes have been studied to better lyse the cells in sludge. Mechanical treatments, which physically disintegrate the cells, include the use of colloid mills (Harrison, 1991), a thickening centrifuge (Michal, 1997), a mechanical jet (Choi, 1997; Kyung, 1997), and freezing and thawing (Chu, 1999). However, anaerobic efficiencies were improved rather lower by mechanical pretreatments which consume a lot of power in comparison with other methods (Weemaes, 1998). Chemical disintegration methods including alkaline and acid hydrolysis (Bien, 1997; Ying, 1997; Lafitte, 2002) have been developed to improve the efficiency of subsequent biological degradation processes. Although the addition of acid or base avoids the necessity of high temperature or high power input, their use is rather limited due to the extreme pH levels required and re-neutralisation after pretreatment.

Biogas production is linked to COD and to the initial biodegradability of the untreated sludge. Thermal treatments are more efficient in terms of solubilisation than other pretreatments including sonication and ozonation (Bougrier, 2008). Therefore, wet air oxidation (WAO) and thermal treatment become potential alternatives in the pretreatment of sewage sludge (Mishra, 1995; Debellefontaine, 1997; Kim, 2003; Yang, 2003). Under mild pretreatment conditions the majority of organic substances in the microbial cell wall are oxidized or pyrolyzed into lower molecular weight compounds and subsequently transferred into the aquatic phase with enhanced biogas production (Yang et al, 2005).

2.1. Effects of Sludge Characteristics on Anaerobic Digestion

2.1.1. Nutrient Contents

Sewage sludge consists of a complex heterogeneous mixture of organic and inorganic materials. The composition and characteristics of sewage sludge depend on several factors like treated wastewater characteristics, sewage sludge treatment process and even the coagulant agents used. Moreover, the wastewater treatment process and municipal and industrial wastewater ratio affect the property of waste sludge. Even if the same wastewater was treated by different process, the concentrations of organic substance, nitrogen, phosphorus and potassium in sludge were quite different. The nutrient contents of sludge in different treatment including conventional activated sludge, oxidation ditch and anaerobic-anoxic-aerobic (A^2/O) are given in Table 2-1.

From Table 2-1, conventional activated sludge shows the highest content in VSS and TN, while sludge from A^2/O has the highest TP and potassium.

Wastewater characteristics are a main influence factor for sludge characteristics. People take plant as the main food in developing countries while take meat and milk in developed countries, and subsequently results in different nutriments in sludge. The municipal sludge in developing countries has its especial characteristics compared with developed countries, given in Table 2-2. Indeed, sludge consists of lower fat and higher carbohydrate (starch, sugar, and fiber) in developed countries and higher carbohydrate in developing countries.

In generally, the order of biogas yield and volatile solid removal is fat> carbohydrate> protein (per unit weight organic substance) during sludge digestion while methane concentrations in biogas is fat > protein>carbohydrate. Therefore, the biogas yield and methane concentration are quite lower in developing countries. The contents of biogas from the anaerobic digestion in some municipal wastewater treatment plants are listed in Table 2-3 in China. The concentrations of CH_4 were about 45-55%.

Table 2.1. The nutrient contents of sludge in different treatment

	Activated sludge	Oxidation ditch	A^2/O
Volatile suspended solids (%)	55	39	37
Total Nitrogen (g/kg)	76	38	43
Total Phosphorus (g/kg)	11	16	24
Total potassium (g/kg)	8	6	14

Wang, 1999.

Table 2.2. Organic contents in sludge in different countries

Nation	Carbohydrate, %	Fat, %	Protein, %	VS Removal, %	Biogas production, $m^3/m^3 \cdot d$	Methane, %
China	>50	20	30	30~50	6~12	45~55.9
German	17	40	43	50~70	17.5	55~56

Yao, 2000; Yang, 2000.

Table 2.3. The contents of biogas from anaerobic digestion in China

CITY	Concentration, %					
	CH_4	CO_2	O_2	H_2	CO	N_2
XI AN	45~53	25~35	3.19	1.79	1.32	9.5
CHENG DU	50	29	1.0		2.5	16.0
AN SHAN	55.9	27.3	1.8	6.9	1.2	8.9
HANG ZHOU	47.7	41.0	0.72			5.7
SHANG HAI	55.1	28.3	0.8	0.27	1.35	11.5

Jin, 1994; Yang, 2000.

Table 2.4. Heavy metal contents in sludge

Treatment plant	Cu	Zn	Pb	Cd	Hg	As	Ni	B	Cr
Beijing Gaobeidian	460.1	790.4	84.5		26	127.7	47.5	20.5	9.2
Shanghai	165.2		110.5	3.6					73.8
Tianjin jizhuangzi	418	1119	295	4.6	4.63	22.5			289
Chongqing Chengnan	69.8	430	90.2	2.72			56		48.5
Hangzhou	576	8696	96	4.78		59	67	58	117
Guangzhou datansha	2200	1790	245				462		1550
Guilin	154	506	199	0.9		37	98		594
Shenyang	55.1	556	155.6	24.1	36.5	12.4			20.5
Xi'an	434	3040	201	695					
Lanzhou		2400			138	560			380

Jin, 1994; Qiao, 2001.
mg/kg.dry.

Transformation and metrology relationships of nitrogen in excessive sludge during the course of anaerobic fermentation were investigated by Dongsu (2008). The excessive sludge was obtained form a sewage treatment plant in which the A/O process was used. The results indicated that TN loss during the course of fermentation were caused mainly by the volatilization of $N-NH_4^+$. While the nitrification and denitrification were contribute to TN loss too. The concentration of $N-NH_4^+$ is correlated positively with the concentrations of VSS and SCOD, but negatively correlated with the concentrations of TN. The metrology relationships indicated that if TN decreased 1.0mg, TSS, TCOD, SCOD and $N-NH_4^+$ decreased -34.0 mg, -68.0 mg, 44.0 mg and 0.40 mg respectively.

2.1.2. Inhibition Contents

1. Heavy Metals

The municipal wastewater is often mixed with industrial wastewater to handle. Most of industrial wastewater contains heavy metals, which results in the complicated composition of sludge and difficulties for disposal. Industrial contributions are the primary source of heavy metals in urban wastewater and account for up to 50% of the total metal content in sewage sludge. Industrial contaminants include zinc, copper, chromium, nickel, cadmium and lead. Domestic sources are mainly associated with leaching from plumbing materials (Cu and Pb), gutters and roofs (Cu and Zn), detergents and washing powders containing Cd, Cu and Zn (Lise, 2008). The concentrations of heavy metals in sludge from different municipal wastewater treatment plants are listed in Table 2-4. Sludge from different municipal wastewater treatment plant contains different heavy metals due to the characteristic of wastewater and treatment process.

The presence of heavy metals not only cause difficulties in the step of the wastewater treatment processes but also influences the sludge disposal due to inhibition. Some metals in trace amounts could be benefit for the activation of many enzymes, but they could cause an inhibitory and even toxic effect on microorganisms in large amounts. The chemical binding of heavy metals to the enzymes and subsequent disruption of the enzyme structure and function are the main cause of this toxic effect. The behavior of heavy metals in wastewater and sludge treatment processes has been widely discussed and summarized in Lisa (2008).

An important feature of heavy metals is that they are not biodegradable and can accumulate to potentially toxic concentrations. The toxic effect of heavy metals is attributed to disruption of enzyme function and structure by binding of the metals with protein molecules or by replacing naturally occurring metals in enzyme groups. (Many heavy metals are part of the essential enzymes that drive numerous anaerobic reactions. Analysis of ten methanogenic strains showed the following order of heavy metal composition in the cell: Fe > > Zn > > Ni > Co = Mo > Cu (Takashima, 1989).

The heavy metal toxicity correlated better to the metal's free ionic concentration than to its total concentration. Because of the complexity of the anaerobic system, heavy metals may be involved in many physic-chemical processes including precipitation or adsorption. The relative sensitivity of acidogenesis and methanogenesis to heavy metals is Cu > Zn > Cr > Cd > Ni > Pb (Lin, 1992) and Cd > Cu > Cr > Zn > Pb > Ni (Lin, 1993). The relative toxicity of four metals to the anaerobic digestion of sewage sludge was reported to be Cr > Ni > Cu > Zn (Industrial wastewaters or sludge generally contain many kinds of heavy metals which cause synergistic or antagonistic effects on anaerobic digestion. The level of inhibition is determined by the species and the ratio of the individual components. Although toxicity of most mixed heavy metals such as Cr–Cd, Cr–Pb, Cr–Cd–Pb, and Zn–Cu–Ni was synergistic (), some of the metal mixtures showed antagonistic inhibition (Lin, 1993). In a variety of aerobic, facultative and anaerobic studies reviewed by Babich and Stotzky (1983), Ni was shown to act synergistically in Ni–Cu, Ni–Mo–Co, and Ni–Hg systems; antagonistically in Ni–Cd, Ni–Zn systems. Ahring and Westermann (1985) found that Ni decreased the toxicity of Cd and Cu.

2. Sulphide

Sulphate is commonly used in industrial and hence in wastewater and waste sludge. Under anaerobic conditions, sulphate is reduced to sulphide by sulphate reducing bacteria (SRB). Sulfate reduction reaction can proceed at appropriate reduction potentials. The reduction potential shows that sulfate is a much less favorable electron acceptor than oxygen (O_2) and nitrate (NO_3^-). In order to maximize the sulfate reduction in wastewater, the reduction potential of the system should be negative. Minimum chemical oxygen demand (COD)-to-sulfate mole ratio of 0.67 is required for achieving theoretically possible removal of sulfate (Choi and Rim, 1991).

Two groups of SRB are responsible for the reduction, the incomplete and the complete oxidizers. In both processes, the reduction half reaction transforms sulphate into sulphur. Inhibition occurs at two different levels. The primary inhibitions caused by the competition for substrates from SRB, and secondary inhibition is due to the toxicity of sulphide for the different groups of microorganisms. Non-dissociated hydrogen sulphide is toxic for both methanogen and sulphate reducers. This form is the toxic form since it can freely diffuse through the cell membrane, causing denaturation of proteins, interfering with the assimilatory metabolism of sulphur.

However, SBR can be used to removal heavy metals. The production of sludge from sulfide precipitation is also low in comparison to hydroxide precipitation. In addition, metal sulfide precipitates are much more stable than metal hydroxide precipitates over a wide pH range. Moreover, valuable metals can be recovered from the metal sulfide sludge (Kaksonen, 2003).

2.2. Pretreatment for Enhanced Anaerobic Digestion

The anaerobic digestion process generally consists of four stages, hydrolysis, acidogenesis, acetogenesis and methanogenesis. In anaerobic digestion, the biological hydrolysis is identified as the rate-limiting step. To reduce the impact of the rate-limiting step, pretreatment of WAS is required such as thermal, alkaline, ultrasonic or mechanical disintegration. These treatments can accelerate the solubilization (hydrolysis) of WAS and reduce the particle size, which subsequently improves the anaerobic digestion.

2.2.1. Mechanical Pretreatment

Mechanical treatments disintegrate the cells and partly solubilize their content by physically strategies. A colloid mill was use for disrupting microbial cells by Harrison (1991). Baier (1997) describes the use of the ball mill and cutting mill for sludge disintegration. In the treatment reactor, moving impellers transfer kinetic energy to grinding glass beads thereby creating high shear stresses that break the cell walls. VS degradation can be improved by 19% and the ball diameter, speed, ball material and sludge concentrations influence the efficiency of anaerobic digestion. Alternative ball mills can be made by ceramic or steel material. Choi (1997) used mechanical jet to enhance the anaerobic digestion. Soluble protein concentration increased by 86% and VSS removal increased 50%.

The mechanical pretreatment for large-scale operation is high-pressure homogenization to compress the sludge. The cells are hereby subjected to turbulence, cavitation and shear stresses, resulting in cell disintegration.

However, it is seen that their efficiency of improving AD of sewage sludge is rather low compared to the other methods. Although most techniques consume a lot of power, they do not require the addition of chemicals or heat.

2.2.2. Chemical Pretreatment

Chemical pretreatment enhance the anaerobic digestion by chemical methods to hydrolyze the cell wall and membrane and thus increase the solubility of the organic matter contained within the cells. Various chemical methods have been developed based on different operating principles including acid and alkaline hydrolysis, advanced oxidation methods.

An acid or base is added to solubilize the sludge, which avoids the necessity of high temperatures. Therefore, the additions of base were widely used to improve the anaerobic digestion. Knezevic (1995) improved gas production with increased NaOH dosage but no significant improvement in VSS reduction was found. Tanaka (1997) studied the addition of NaOH with heating on the effect of anaerobic digestion. Biogas production was increased by 20% with 50% improvement of methane production. The methods are shown to be an effective for sludge solubilisation. As required extreme pH levels and re-neutralization after pretreatment, the chemical pre-treatment for anaerobic digestion is rather limited.

2.2.3. Thermal Pretreatment

Biogas production is linked to COD and to the initial biodegradability of the untreated sludge. Thermal treatments are more efficient in terms of solubilisation than other pretreatments including sonication and ozonation (Bougrier et al, 2008). Therefore, wet air oxidation (WAO) and thermal treatment become potential alternatives in the pretreatment of sewage sludge (Mishra et al 1995; Debellefontaine, 1997; Kim et al, 2003; Yang et al, 2003). Under mild pretreatment conditions the majority of organic substances in the microbial cell wall are oxidized or pyrolyzed into lower molecular weight compounds and subsequently transferred into the aquatic phase with enhanced biogas production (Yang et al, 2005).

2.2.4. Ultrasound Pretreatment

Microwave (MW) pretreatment is an alternative method to destroy pathogens and to increase VS destruction (Decareau, 1985). MW irradiation can produce focused direct heat rapidly and subsequently cause the breakage of cell walls and denaturing of complex biological molecules. MW effect on WAS solubilisation and improvements in VS destruction has recently been reported Eskicioglu (2007) and Coelho(2008). Park (2004) studied the effect of MW pretreatment, which improved biogas production 36%.

Although cell disintegration can be successfully obtained at high power levels, the high power consumption becomes a serious question.

2.2.5. Advanced Oxidation Pretreatment

Oxidants including ozone and peroxide is used for the destruction of pathogens and cell walls in WAS based on the generation of hydroxyl radicals. Moreover, hazardous by-products were not detected. Although oxidative treatments are considered promising, additional research is needed to avoid extreme reaction conditions in terms of pressures and temperatures, or pH.

2.3. THERMAL PRETREATMENT KINETICS

Sludge WAO becomes potential alternatives to pretreat waste sludge. Under a mild condition of wet air oxidization, the majority of organic substance that enwrapped by microbial cell wall oxidizes into small molecule and shift into aqueous phase and thus enhances the biogas production. Therefore, it is necessary to investigate the mechanisms of WAO in order to achieve a better understanding of the reaction process and get a better reactor design (Rivas, 1998).

2.3.1. Kinetic Model

Takamatsu et al (1970) suggested a model for thermal treatment of activated sewage sludge. They assumed the sludge to be consisting of solid matter and soluble nonevaporative matter; soluble evaporative matter and water. Foussard (Foussard, 1989) has ever divided the sewage sludge in two parts, easily oxidizable and difficult to oxidizable. They observed that the easily oxidizable was about 60% and the difficult to oxidizable was 40%. Reaction rate coefficient calculated for the easily oxidizable was closer to 2-butanol while that for the difficult oxidizable was close to sodium acetate oxidation. The point selectivity (α) for activated sludge was 0.15(Li, 1991), but it was 1 on the basis of data of Foussard et al. One reason for this contradictory observation may be due to the different parameters measured by Ploos Van Amstel and Rietema (COD) and Foussard (TOC) to follow the WAO reaction. Another reason may be that two-stage first-order reaction model represents well the WAO treatment process of many phenolic wastewaters(Portela, 1997; Kolaczkowski, 1997) but not sludge. However, they all perceived the resistance of the acetic acid to further oxidation.

Indeed, reaction steps in sludge WAO can be included as follows: ① oxygen transfer, oxygen transfer from the gas phase to liquid phase and reaches to the surface of sludge; ② Oxidization reaction, oxygen react with organics clinging to the surface of sludge and cell walls in sludge and then the microbial cell walls split and substances enwrapped in the cell wall disperse into the liquid phase; ③ Oxidization reaction, substances are oxidized into volatile fatty acid, amino acid and CO_2.

Due to fast transport of oxygen in the gas phase and slightly soluble oxygen in water as well as the forceful stirring in reactor, diffusion resistance can be ignored. Therefore, the oxidization reaction can be considered as the limitation step in the whole WAO process. The MLVSS/MLSS ratio was selected as parameters of model, so the data can't be affected by the weight of samples and liquid and solid ratio like COD or TOC.

2.3.2. The Two-stage First-order Reaction Kinetic Model

Assuming the WAO oxidation of the original organic compounds proceeds in two steps, the organic compounds are converted into small-molecule intermediates in the first step of the WAO reaction. Those intermediates are subsequently reduced to final products in the second step. The reaction rate coefficient is related to the reaction temperature which comply with Arrhenius equation. The first order kinetic model is given

$$\frac{d[c]}{dt} = k_0 \exp\left(-E_a / RT\right) [c]^m [O]^n \qquad (1)$$

In which c and c_0 are respectively the initial MLVSS/MLSS concentration and that at time t. where k_0 the pre-exponential factor, E_a activation energy, R the gas constant, and T the temperature. Generally, n is very small or oxygen is excessive (n=0), so the item of oxygen content $[O]$ is a constant. Equation [1] can be simplified and integrated as follows:

$$\ln\left(\frac{c}{c_0}\right) = k_0 \exp(\frac{-Ea}{RT})t \qquad (2)$$

Equation [2] is the basic form of first-order reaction kinetic model of wet air oxidization process, while Ea and k_0 can be calculated according to this model.

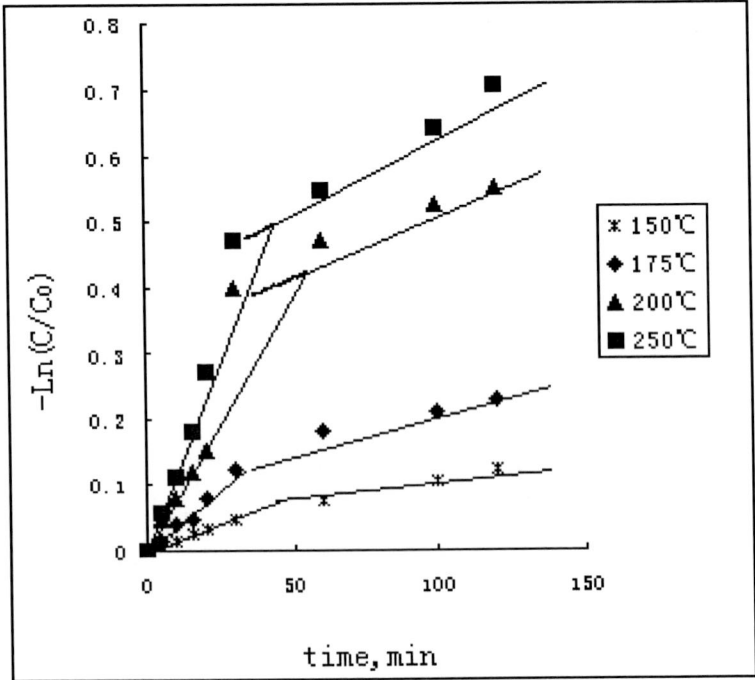

Figure 2.1. Petrochemical sludge of calculated profiles and experimental data.

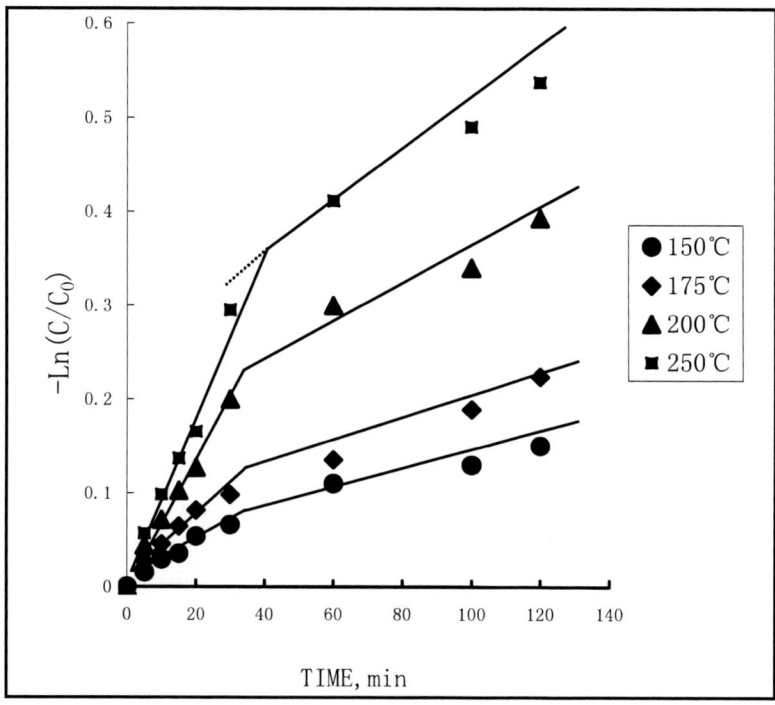

Figure 2.2. Refinery oil sludge of calculated profiles and experimental data.

Table 2.5. Rate coefficient of two-stage first-order reaction kinetic model

Temperature °C	Petrochemical sludge				Refinery sludge			
	K_1, min^{-1}	R^2	K_2, min^{-1}	R^2	K_1 min^{-1}	R^2	K_2 min^{-1}	R^2
150	0.0016	0.999	0.0008	0.995	0.0027	0.984	0.0009	0.952
175	0.0037	0.971	0.001	0.948	0.0043	0.983	0.0014	0.996
200	0.0075	0.997	0.0016	0.974	0.0067	0.978	0.002	0.947
225	0.0126	0.985	0.0026	0.995	0.0089	0.971	0.0026	0.976

Joglekar(Joglekar HSetal.,1991) even considered that there were three types in this model: a. the fast-reaction phase after slow-speed leading phase; b. no leading phase; c. the slow-reaction phase after fast-reaction phase .In types of a and c, the same reactive process has two different reaction rate coefficient, why is so-called kinetic model of two-stage first-order reaction.

Table 2.6. Activation energy and pre-exponential factor on two-stage first-order reaction kinetic model

	First step			Second step		
	Ea, kJ/mol	k_1^0	R^2	Ea, kJ/mol	k_2^0	R^2
Petrochemical sludge	48.533	1142	0.997	21.408	0.3497	0.996
Refinery sludge	28.259	8.468	0.999	24.833	1.090	0.996

Figure2-1 and Figure2-2 show the first-order kinetic model fit of the total removal of MLVSS/MLSS in the wet air oxidization process for petrochemical sludge and refinery sludge, respectively. The dispersed points in figure are experimental data and the line is obtained by least square fitting.

The kinetic rate coefficient for each oxidation step was obtained and listed in Table 2-5.

According to the reaction rate coefficient K under different temperature in Table 3, LnK and $1/T$ are used to make a drawing in which slope is $-Ea/R$ and the intercept is Lnk_0, activated energies and pre-exponential factors are listed in Table 2-6. It's apparent that the rate coefficients follow the Arrhenius correlation well.

From Table 2-6, it shows $K_1 > K_2$, namely both the wet air oxidization processes of waste sludge follow the slow-reaction stage after fast-reaction stage. However, activation energy and the pre-exponential factor are quite different in different waste sludge WAO process.

2.3.3. Generalized Kinetic Model

In the generalized model (Li,), there are three groups of organic substances: initial compounds, refractory intermediates, oxidation end products. This model was based on the assumption that some of the organic compounds present in the wastewater are directly oxidized to the final products, while the other is first converted to an intermediate product, which is then further oxidized to the final products.

In this model, three groups of organic substances are defined to exist in the liquid and gas. Group A: initial compounds and relatively unstable intermediates; group B: refractory intermediates; group C: final oxidation products. The wet air oxidation process could then be expressed as follows:

Potential of Anaerobic Sludge Digestion

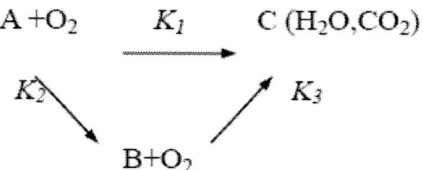

Assuming all the reactions above are first-order reactions and the rate coefficient is related to the reaction temperature according to the Arrhenius equation. Therefore, the generalized kinetic model can be given by :

$$-\frac{d[A]}{dt} = (K_1 + K_2)[A] \tag{3}$$

$$-\frac{d[B]}{dt} = K_3[B] - K_2[A] \tag{4}$$

where K_1, K_2 and K_3 are :

$$K_1 = k_1^0 e^{-E_1/RT}[O_2]^{n_1}, \quad K_2 = k_2^0 e^{-E_2/RT}[O_2]^{n_2},$$
$$K_3 = k_3^0 e^{-E_3/RT}[O_2]^{n_3}$$

As oxygen is excessive in reaction, while A_0 is the initial concentration of group A and B_0 is the initial concentration of group B, namely, $[A] = [A_0]$, $[B_0] = 0$

The total removal of MLVSS/MLSS at any time is:

$$\frac{[A+B]}{[A+B]_0} = \frac{K_2}{K_1 + K_2 - K_3} e^{-K_3 \cdot t} + \frac{K_1 - K_3}{K_1 + K_2 - K_3} e^{-(K_1 + K_2) \cdot t} \tag{5}$$

The ratio of the formation rate coefficient of group B to group C is defined as the point selectivity $\alpha = K_2/K_1$, where α can be considered as the completion level of wet air oxidization. If there are many short-chain organic substances in WAO, the value of α could be high.

Table 2.7. Rate coefficients of generalized reaction kinetic model

	Temperature °C	Rate coefficient of generalized reaction kinetic model			
		K_1(min^{-1})	K_2(min^{-1})	K_3(min^{-1})	α
Petrochemical sludge	150	0.00274	0.0168	0.00011	6.13
	175	0.00561	0.0252	0.00042	4.49
	200	0.00938	0.0336	0.0020	3.58
	225	0.01197	0.0372	0.0030	3.11
Refinery sludge	150	0.00280	0.0172	0.000095	6.14
	175	0.00489	0.0262	0.00054	5.36
	200	0.00877	0.0333	0.00181	3.80
	225	0.01210	0.0387	0.00284	3.20

Table 2.8. Activation energy and pre-exponential factor of generalized reaction kinetic model

	A→C			A→B			B→C		
	Ea, kJ/mol	k_1^0	R^2	Ea, kJ/mol	k_2^0	R^2	Ea, kJ/mol	k_3^0	R^2
Petrochemical sludge	34.948	59.06	0.96	18.897	3.91	0.97	80.761	1.12×10^6	0.97
Refinery sludge	34.965	61.64	0.97	18.919	3.85	0.99	80.963	1.20×10^6	0.96

Origin7.0 can be used to fit the experimental data by generalized kinetic model. The rate coefficients and the point selectivity were listed in Table 2-7 as a function of temperature, where clearly shows a strong temperature dependence of the kinetic parameters. K_1, K_2, and K_3 increase significantly with an increase in temperature.

According to Arrhenius equation, activated energy and pre-exponential factor can be calculated and listed in Table 2-8. Both of the values of activation energy and pre-exponential factor of waste sludge are very close. Another important point to note in Table 2-7 is the value of α which is larger than 3 for all of the cases studies here. In addition, at the same temperature, α of refinery sludge is slightly higher than that of petrochemical sludge. Those imply that there is a strong presence of acetic acid in the present WAO treatment process and the concentrations of acetic acid and short chain organic substance in the supernatant of refinery sludge are higher than that of petrochemical sludge.

Table 2.9. Components of supernatant of waste sludge after WAO treatment by chromatogram-mass spectrograph instrument（mg/L）

	Petrochemical sludge	Refinery sludge		Petrochemical sludge	Refinery sludge
Formic acid	23	58	Acetaldehyde	15	10
Acetic acid	1230	396	Propylaldehyde	4	2
Propanedioic acid	760	116	Butyraldehyde	13	6
Butyric acid	76	13	3-Methyl-butanal	9	4
Methacrylic acid	75	39	Benzal dehude	8	7
3-methyl-butyrate	190	70	5- Methyl-furfural	19	68
3-butenoic acid	158	17	Methanol	11	5
2-butenoic acid	700	118	Ethanol	7	3
Eicosapentaenoic Acid	36	4	Furfuryl alcohol	380	130
Heptenoic acid	37	13	2-ethyl hexanol	5	20
4-pentenoic acid	532	38	5- methylic Butyrolactone	40	13
2-ethylhexoic acid	21	27	Butyrolactone	39	14
Formamide	70	23	Carbolic acid	221	36
Acetamide	1118	475	Isobutyramide	75	27
Acrylamide	35	26	Valeramide + Hexenic acid	265	46
concentration of acetic acid,%	19.9	21.7	concentration of short chain organic substance, (includes 1 or 2 carbon atoms), %	40.1	53.2

Those were experimentally confirmed by chromatogram-mass spectrograph instrument. Table 2-9 list components of the supernatant of waste sludge after the wet air oxidation process at 200°C. The concentrations of acetic acid and short chain organic substance in the supernatant of refinery sludge are 21.7% and 53.2% respectively, but those of petrochemical sludge are 19.9% and 40.1% respectively. These results are coincident with α obtained from generalized kinetic model.

Two-stage first-order reaction kinetic model used to fit sludge WAO process show different values of activation energy and the pre-exponential factor. WAO process of industrial waste sludge has been shown to follow reasonably well the generalized kinetic model. The values of α, point-selectivity, predict a strong presence of acetic acid and short chain organic substance in the WAO treatment process.

2.4. Energy Output Potential

Kim (2003, Effects of Various Pretreatments for Enhanced Anaerobic Digestion with Waste Activated Sludge) was to investigate the effect of various pretreatments (thermal, chemical, ultrasonic and thermochemical pretreatments) of WAS on solubilization, particle size reduction and methane production enhancement.

Biogas production following the anaerobic digestion of the thermally (4843 l/m^3WAS), chemically (4147l/m^3WAS), ultrasonically (4413l/m^3WAS) and thermochemically (5037 l/m^3WAS) pretreated WAS was higher than that of traditional anaerobic digestion (3657 l/mWAS). Methane production levels also obtained by the anaerobic digestion of the thermally (3390 l/m^3WAS), chemically (2827 l/m^3WAS), ultrasonically (3007l/m^3WAS) and thermochemically (3367 l/m^3WAS) pretreated WAS were higher than that of the control (2507 l/m^3WAS). In conclusion, thermal and thermochemical pretreatment methods showed good results.

COD removal and VSS removal were enhanced and higher methane production was obtained by pretreatment. However, pretreatment processes lead to more energy input. Therefore, the energies of input and output are discussed in order to gain a better understanding of the energy balance.

The energy required to heat the sludge can be assessed by considering the sludge heat capacity equal to the water heat capacity (Bougrier et al, 2006). Energy requirement really depends on the required pretreatment temperature. Indeed, the most energy required is utilized in water vaporization (2260 kJ/kg) when the pretreatment temperature is above 100°C. Although sludge heating costs can be lowered by working with higher feed sludge concentrations, the thickened sludge could deposit in the heating exchanger resulting in lower efficiency of the heat exchanger and even block the pipeline or the heat exchanger. However, the heat exchanger could considerably reduce the heating energy requirement. If the efficiency of the heat exchanger is considered to be 95%, the total input energies are 135.1 kJ/kg at WAO 150°C and 140.3 kJ/kg at WAO 200°C, without regard to the heat output from sludge oxidation or sludge pyrolysis, given in Table 5.

Considering the energy obtained from the biogas production, the results indicate that the energy output in the whole process decreased with increasing pretreatment temperature and that net energy outputs were positive above 100°C when the efficiency of the heat exchanger is above 97%.

In the WAO-UASB process, the concentrations of acetic acid and volatile fatty acids as well as BOD/COD rates in the supernate increased with increasing pretreatment temperature. Biodegradability of the supernate and filtration properties of the residue were significantly improved. However, total N concentrations were still higher in the final effluent and lower in the final residue after pretreatment by the UASB process than after traditional anaerobic digestion. More than 90% of the TN was removed by anaerobic ammonium oxidation in the anaerobic digestion stage.

Pretreatment-UASB process showed much lower concentrations of TP in the final effluent and higher TP content in the residue. Also, most metals remained condensed in the solid residue (Al, Ca, Fe, Ni, Cu, Zn, and Ba). In contrast, Mg and Na were mainly present in the aquatic phase. Total sulphur in residues decreased.

In terms of COD added, biogas production was 2.78 times greater after the 200°C WAO pretreatment compared with the traditional anaerobic digestion. However, in terms of added raw sludge, pretreatment had no significant effect on biogas production. Comparing the energy obtained from biogas production with energy inputs required for pretreatment, showed that energy output for the whole process decreased with increasing pretreatment temperature and that net energy outputs were positive above 100°C when the efficiency of the heat exchanger is above 97%.

ABBREVIATIONS

BOD, biochemical oxygen demand (5 days, 20°C) (mg O_2/l); COD, chemical oxygen demand (mg O2/l); HRT, hydraulic retention time (d); CODs, soluble fraction of the chemical oxygen demand (mg O_2/l); VSS, volatile suspended solids; MLVSS, mixed liquid volatile suspended solids; WAS, waste-activated sludge; TP, total phosphorus; TN, total nitrogen; TS, total sulphur.

REFERENCES

Bien, J.B., Kempa, E.S., Bien, J.D., 1997. Influence of ultrasonic field on structure and parameters of sewage sludge for dewatering process. *Water Sci. Technol.*, 36 (4), 287–291.

Bougrier, C., Delgenes, J.P., Carrere, H., 2006. Combination of thermal treatments and anaerobic digestion to reduce sewage sludge quantity and improve biogas yield. *Process Saf. Environ. Prot.*, 84(B4), 280–284.

Bougrier, C., Delgenes, J.P., Carrere, H., 2007. Impact of thermal pre-treatments on the semi-continuous anaerobic digestion of waste activated sludge. *Biochem. Eng. J.* 34, 20–27.

Bougrier, C., Delgenes, J.P., Carrere, H., 2008. Effects of thermal treatments on five different waste activated sludge samples solubilization, physical properties and anaerobic digestion. *Chem. Eng. J.* 139 (2), 236–244.

Canales, A., Parallax, A., Rols, J. L., Goma, G. Huyard, A., 1994. Decreased sludge production strategy for domestic wastewater treatment. *Wat. Sci. Tech.* 30, 97-106

Cao Renlin et al. 1997. Research on effect of sewage sludge compost applied for green garden. *Environment Science Research*. 3, 46~50

Choi HB, Hwang KY, Shin EB. 1997. Effects on anaerobic digestion of waste activated sludge pre-treatment. *Water Sci. Technol.* 35,207–11.

Chu, C.P, Feng, W.H., Chang, B.V., Chou, C.H., Lee, D.J., 1999. Reduction of microbial density level in wastewater activated sludge via freezing and thawing. *Water Res.* 33(16), 3532-3535.

Cutforth. S.J., 1995. Preliminary results of the anaerobic biotreatability of the effluent from the wet-air oxidation of sewage sludge. *Journal of the Chartered Institution of Water and Environment Management* 9, 231-235.

Debellefontaine H., Cammas F. Deiber X., G., Foussard J. N., Reihac P., 1997. Wet air oxidation: kinetics of reaction, carbon dioxide equilibrium and reaction design- an overview. *Water Sci. Technol.*, 35(4), 111-118.

E. Choi and J.M. Rim, 1991. Competition and inhibition of sulfate reducers and methane producers in anaerobic treatment. *Water Sci Technol.* 23, 1259–1264.

F. J. Rivas, S.T. Kolaczkowski, et al, 1998. Development of a model for the wet air oxidation of phenol based on a free radical mechanism, *Chem. Eng. Sci.* 14, 2575-2586.

Fen Wang, Yong Wang, Min Ji, 2005. Mechanisms and kinetics models for ultrasonic waste activated sludge disintegration. *J. Hazard. Mater.* 123, 145-150.

Foussard, J. N., et al., 1989. Efficient elimination of organic liquid wastes. wet air oxidation. *Environ. Eng.* 115, 367~385

H. Debellefontaine, F. X. Cammas, G. Deiber, J. N. Foussard and P. Reihac, 1997. Wet air oxidation: kinetcs of reaction, carbon dioxide equilibrium and reaction design- an overview. *Wat. Sci. Tech.*, 35, 111-118

Harrison, S.T.L., 1991. Bacterial cell disruption: a key uint opertion in the recovery of intracellular products. *Biotech. Adv.*, 9, 217~240

Harrison, S.T.L., 1991. Bacterial cell disruption: a key unit operation in the recovery of intracellular products. *Biotech. Adv.,* 9,217~240.

Jeong Tae-Young, Cha Gi-Cheol, Seo Yung-Chil, Jeon Choong, Suk Soon Cho, 2008. Effect of COD/sulfate term ratios on batch previous anaerobic digestion using waste activated sludge. *J. Ind. Eng. Chem.* 14(5), 693-697.

Joglekar H. 1991. Kinetics of wet air oxidation of phenol and substituted phenols. *Wat.Res.* 2,135~145

Julia Kopp, Johannes Muller, Norbert Dichtl, Jörg Schwedes, 1997. Anaerobic digestion and dewatering characteristics of mechanically disintegrated excess sludge. *Water Sci. Technol.* 36(11), 129-136.

Kim Jeongsik, Chulhwan P, Tak-hyun K, Myunggu L, Sangyong K., 2003. Seung-wook K, Linwon L., Effects of various pretreatments for enhanced anaerobic digestion with waste activated sludge, *J. Biosci. Bioeng.*, 95,271-275.

Kolaczkowski, S.T.; Beltran, F.J.et al.,1997. Wet air oxidation of phenol: factors that may influence global kinetics. Process Safety and Environmental Protection: Transactions of the Institution of Chemical Engineers, Part B. 4, 257-265

KyungYub Hwang, EungBai Shin, HongBook Choi, 1997. A Mechanical pretreatment of waste activated sludge for improvement of anaerobic digestion system. *Water Sci.Technol.* 36(12), 111~116.

Kyung-Yub Hwang,Eung-Bai Shin,Hong-Book Choi, 1997. A Mechanical pretreatment of waste activated sludge for improvement of anaerobic digestion system. *Wat.Sci.Tech.* 12, 111~116

Lafitte Trouqure S., Forster, C.F., 2002. The use of ultrasound and irradiation as pre-treatments for the anaerobic digestion of waste activated sludge at mesophilic and thermophilic temperatures. *Bioresour. Technol.* 84, 113–118.

Li, L., Chen, P., Gloyna,E.F., 1991. Generalized kinetics model for wet air oxidation. *AICHE J.* 37, 1687-1697

Lise Appels, Jan Baeyens, Jan Degrève, Raf Dewil, 2008. Principles and potential of the anaerobic digestion of waste-activated sludge. Progress in Energy and Combustion Science 34(6), 755-781.

Lopesa S.I.C., Dreissena C., Capelab M.I., Lens P.N.L., 2008. Comparison of CSTR and UASB reactor configuration for the treatment of sulfate rich wastewaters under acidifying conditions. *Enzyme Microb. Technol.* 43, 471–479.

Michal Dohanyos, Jana Zabranska, Pavel Jenicek, 1997. Enhancement of sludge anaerobic digestion by using of a special thicking centrifuge. Water Sci.Technol. 36(11), 145~153.

Mishra V. S., Joshi J. B., Mahajani V. V., 1994. Kinetics of wet air oxidation of diethanolamine and morpholine. *Water Res.* 28(7), 1601~1608.

Mishra, V. S., Mahajani, V. V. and Joshi, J. B., 1995. Wet air oxidation. *Ind. Engng. Chem. Res.* 34, 2-48

Ploos Van Amstel, J. J. A.; Rietema, K. 1973. Wet air oxidation of sewage sludge. *Chem. Eng. Tech.* 20, 1205-1211

Portela Miguelez, J.R. Lopez Bernal, J., et al., 1997. Kinetics of wet air oxidation of phenol. *Chemical Engineering Journal*, 2, 115-121

Qiao Xianliang, Luo Yongming, Wu Shengchun, 2000. Application in agriculture and its affects of sewage sludge. *Soil* 2, 79~85

S. Tanaka, T. Kobayashi, K. Kamiyama, M.L. Signey Bildan, Effects of thermochemical pre-treatment on the anaerobic digestion of waste activated sludge, *Water Sci. Technol.* 35 (1997), pp. 209–215.

Takamatsu, T., et al., 1970. Presented at 5th Int. Water pollut. Res.Conf. paper II-32

Tufano, V, 1993. The multi-step kinetic model for phenol oxidation in high pressure water. *Chem. Engng. Technol.* 16, 186~190

U. Baier, P. Schmidheiny, 1997. Enhances anaerobic degradation of mechanically disintegrated sludge, *Water Sci. Technol.* 37, 137–143.

Wang Hongkang,Yan Shoucang, 1990. Research of crop pollution of Cu for applied sewage sludge. *Enviroment Science* 3, 6-11

Weemaes MPJ, Verstraete W., 1998. Evaluation of current wet sludge disintegration techniques. *J. Chem. Technol. Biotechnol.* 73,83–92.

Wild D., Kisliakova A., Siegrist H., 1996. P-fixation by Mg, Ca and zeolite A during stabilization of excess sludge from enhanced biological P-removal. *Water Sci. Technol.* 34(1-2), 391-398.

Xiaoyi Yang, Zhanpeng Jiang et al, 2003. Study on Treatment of waste sludge by wet air oxidation and two-phase anaerobic process. *China Water and Wastewater* 4, 48-51

Xiaoyi Yang, Zhanpeng Jiang et al. 2005. Study on two-phase anaerobic treatment of waste sludge after wet air oxidation treatment, water and wastewater 1, 54-58

Yang Xiaoyi, Jiang Zhanpeng, 2003. Study on Treatment of sewage sludge by wet air oxidation. *Water and Wastewater Engineering* 29(7), 50-54.

Yang Xiaoyi, Jiang zhanpeng, 2005. Study on two-phase anaerobic treatment of waste sludge after wet air oxidation treatment. *Water and Wastewater Engineering* 31(1), 54-58

Yingchih Chiu, Chengnan Chang, Jihgaw Lin, Shwujiuan Huang, 1997. Alkaline and ultrasonic pretreatment of sludge before anaerobic digestion. *Water Sci. Technol.* 36(11), 155-162.

Z. Knezevic, D.S. Mavinic and B.C. Anderson, Pilot scale evaluation of anaerobic codigestion of primary and pretreated waste activated sludge, *Water Environ. Res.* 67 (1995), pp. 835–841.

Chapter 3

3. POTENTIAL OF SLUDGE PYROLYSIS

3.1. PRINCIPLES

Pyrolysis is the chemical decomposition of a substance by heating, which pyrolysis is a special case of thermolysis in the absence of oxygen. Sludge pyrolysis involves the thermal decomposition of sewage sludge in the absence of oxygen.

Thermal techniques offer an interesting alternative for the stabilization or elimination of sewage sludge. Moreover, pyrolysis offers the possibility of recovering a source of chemicals or fuels, while safely enclosing heavy metals in the solid residues that are formed during the reactions. Pyrolysis systems should have operating and construction costs similar to those of sludge combustion systems while providing the added benefits of converting sewage sludge to energy and chemicals, which should make sludge pyrolysis profitable. Furthermore, the residual solids formed by the pyrolysis of sewage sludge under controlled conditions can be used as an adsorbent for the control of air pollution (Ros et al, 2006).

Moreover, particular interest is shown in this process as a high recovery of liquid oil is achieved, lower emissions of NO_x and SO_x also lower operating costs when compared to incineration. Bayer (1981) designed a continuous lab-scale plant that demonstrated the production of synthetic crude oil under oxygen free conditions. The operating conditions were around 300°C, at atmospheric pressure with a residence time of three hours. The fundamental mechanics of the conversion reactions proved that organic substrates generate liquid hydrocarbons by thermal decomposition with metal catalysts and the proteins and lipids in sewage sludge generate oil and char.

Results have shown that the oil obtained from sewage sludge pyrolysis can be used directly in diesel fuelled engines and is comparable to low-grade petroleum distillates from commercial refineries (Lilly, 2003).

3.2. Affecting Parameters

Three products are usually produced: oil, char and gas in the sludge pyrolysis. The relative proportions of these depend on the pyrolysis methods and reaction parameters such as final temperature, heating rate, residence time and sample size.

3.2.1. Temperature

The temperature is an important factor for sludge pyrolysis. Organic bonds have been broken at different temperatures during the pyrolysis. At the temperature range of 500–575 °C, carboxylic, phenolic and ether oxygen are broken. In general, lower process temperature and longer vapour residence time favour the production of char. High temperature and longer residence time increase the biomass conversion to gas and moderate temperature and short vapour residence time are optimum for producing liquids. Above 600 °C most of the carboxylic and phenolic bonds have been broken while cellulosic break up to 650 °C. The retention time of the pyrolysis reaction decreases with the rise of the temperature. Moreover, the temperature doesn't only affect the product yields, but also affect the heat value of the products. The calorific value of oil and char often decrease with the rise of the temperature.

Fast pyrolysis for liquids production is of particular interest currently as the liquids are transportable and storage. Lilly (2003) studied that the oil recovery from sewage sludge by low-temperature pyrolysis in a fluidized-bed. The oil yields increase with increasing temperature initially as sludge is subjected to more energy, stronger bonds break and thus an increase in larger compounds are observed. The decrease in oil yields above 525 °C is believed to be a result of secondary decomposition reactions which break the oil into lighter, gaseous hydrocarbons. As a consequence the NCG (a non-condensable gas) yields also increase. Char yields are expected to decline as more volatiles are released.

3.2.2. Gas Residence Time

The effect of gas residence time influences the trends of product yields. At a constant pyrolysis temperature, oil and char yields are both at maximum at the shortest residence time, which is in agreement with a variety of biomass and coal feeds. The decrease of char yield is accompanied by evolution of more volatile matter from the sewage sludge at the longest residence time.

If the pyrolysis vapor stays at the reactor more time, some oil will change into gas with a result of secondary cracking. The short gas residence time often results to the largest oil yield and the smallest NCG yield.

3.2.3. Heating Rate and Feed Size

Comparing with pyrolysis temperature, heating rate is not very important. However, heating rate is very important for fast pyrolysis. The effect of the heating rate was found to be important only at low final pyrolysis temperatures. The high heat rate increases the oil yield and decreases the specific surface area of char.

There exits an optional feed size to make the oil yield biggest and the smaller the particle size, pyrolysis temperature lower for the biggest oil yield.

Table 3.1. Effects of feed size and pyrolysis temperature on oil yield

Feed size (mm)	Oil yield (%)	Pyrolysis temperature (°C)
1~0.5	29.3	500
0.5 ~ 0.2	32	450
<0.2	30.7	400

3.2.4. Sludge Type

Sewage sludge is the waste produced in the wastewater treatment plants. Its composition may change due to the origin and to the non-standardized treatments in the wastewater treatment plants. The biggest distinction is the concentration of volatile because much of oil comes from it. The increase of the volatile content in the sewage sludge samples caused an increase in the liquid yield. In general, the oil yield is around 30% by activated sludge pyrolysis while that is around 10% from digestion sludge pyrolysis.

3.3. Pyrolysis Products

3.3.1. Oil

The oil obtained from sludge pyrolysis has a high viscosity and high heating values of 29 - 38 MJ/kg and can be used as fuel. The elemental characteristics of the oil produced were quiet stable over a wide of operating conditions. The typical ratio of C: H: O: N: S was 76: 11: 6.5: 4: 0.5 in terms of its carbon content. It corresponds to that of heavy crude oil.

Park (2008,) obtained oils with high caloric value and low contaminant content by pyrolysis of digested and dried sewage sludge containing polymer flocculants, which was pyrolyzed in a temperature range of 400 – 700 °C in a pyrolysis plant equipped with a fluidized bed reactor. Above 50 wt% pyrolysis oil was obtained with a maximum caloric value of 33 MJ/kg. By using a char separation system composed of a cyclone and a hot filter and calcium oxide as an adsorbent, the pyrolysis oil was almost free of hazardous metals and its chlorine content was significantly reduced.

3.3.2. Char

The chars from sludge pyrolysis can be used as fuel, fillers and adsorbent. Especially, the chars from pyrolysis as adsorbent for adsorption become to attract lots of studies. Specific surface area is an important characteristic. The benefits of this technology include a reaction in the volume of the sewage and the production of a cheaper adsorbent than commercial activated carbon. The elemental characteristics of the char were affected by operating condition. The carbon, hydrogen and nitrogen contents in the char often decrease with increased temperature.

The weight loss of the sludge sample is dependent on temperature. The pore structure of char as adsorbent was created due to removal of volatile matter, and temperature is a major factor in enhancing the surface area (Lu, 1995, 1996).

He also discussed the characteristics of adsorbents derived from sewage sludge by chemical activation methods. Zinc chloride degrades cellulose materials and upon heating, it causes dehydration leading to charring and aromatization of the carbon skeleton and creation of pore structure. Optimum activating temperature is found to be 650°C with zinc chloride with BET surface area (309 m^2/g) and micropore area (171 m^2/g). At 650°C, the optimal zinc chloride concentration is 5 M giving highest BET surface area. For micropore area, 3 M gives the best results.

Experimental results show clearly that the sludge derived adsorbent is capable of removing H_2S by adsorption. Its capacity is about 25% that of the commercial activated carbon, under the laboratory test conditions used.

Sewage sludge(Charothon, 2007) can be used to develop a potential adsorbent for dye removal by pyrolysis under either N_2 or CO_2 atmospheres. The surface area of the char increases as the pyrolysis temperature increase under the CO_2 atmosphere. The maximum surface area of the char is achieved with pyrolysis at 750 °C under the CO_2 atmosphere with 60.7 m^2/g mainly mesoporous. The FT-IR spectra of the char prepared under both N_2 and CO_2 atmospheres indicate a decrease in –OH, –NH and C=O functionalities with increasing the pyrolysis temperature, corresponding to a decrease in the acidity of the char. The adsorption mechanism is governed by the combination of the electrostatic interactions and dispersive interactions. The equilibrium data fit well with the Langmuir model of adsorption suggesting a monolayer coverage of dye molecules at the outer surface of sewage sludge derived chars.

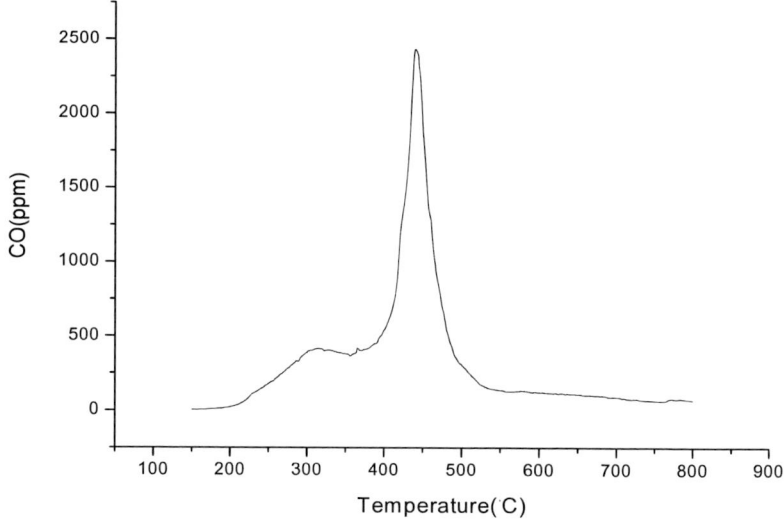

Figure 3.1. Carbon monoxide concentration discharged during sludge pyrolysis.

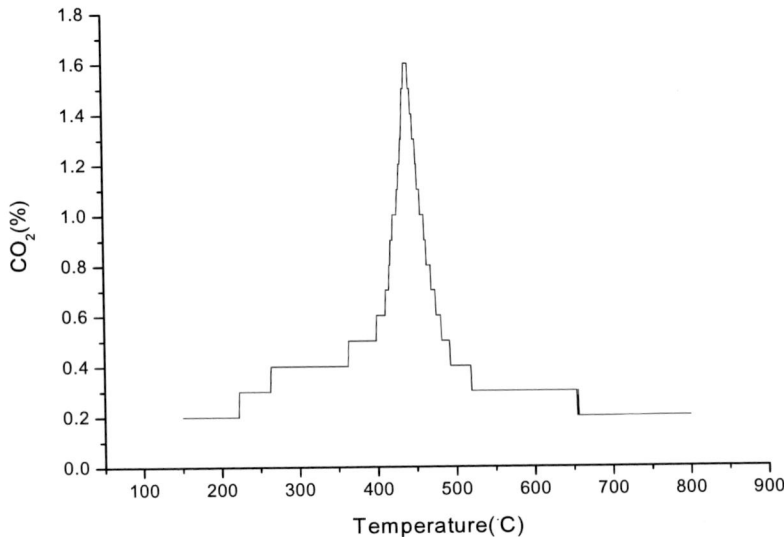

Figure 3.2. Carbon dioxide concentration discharged during sludge pyrolysis.

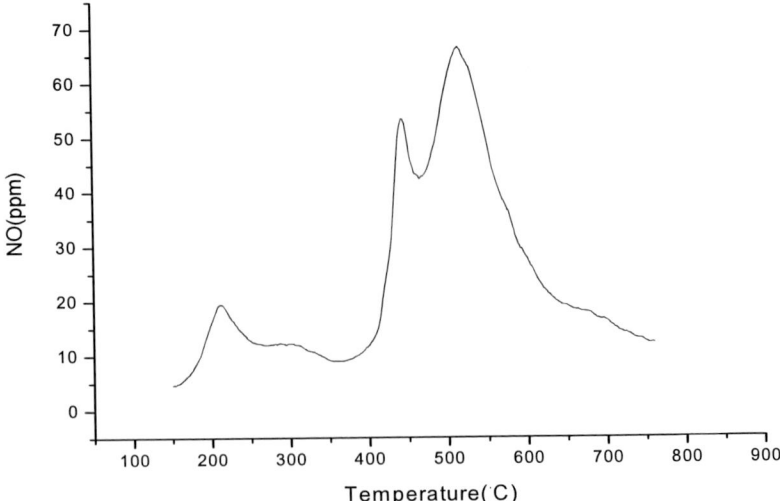

Figure 3.3. Nitrogen oxide concentration discharged during sludge pyrolysis.

3.3.3. Non-condensable Gas

The components of pyrolysis gas are carbon dioxide, carbon monoxide, nitrogen oxide. The main component is carbon dioxide at range of 200 – 600°C, given in Figure 3-1, Figure 3-2, Figure 3-3, while the concentration of carbon monoxide discharged at the same temperature range as carbon dioxide did. The concentration of nitrogen oxide changed during the whole temperature range in sludge pyrolysis.

3.4. PRETREATMENT FOR SLUDGE PYROLYSIS

3.4.1. Acid and Base Pre-treatment

The use of acidic pretreatment has been employed to enhance the adsorptive properties of chars that were generated by sludge pyrolysis (Rio et al., 2005). The surface area of adsorbent produced from pyrolyzed sludge increased with acid treatment (Lu and Lau, 1996). Moreover, pretreatment results in enhanced oil production during pyrolysis by modification of the structure of sludge-based organic matter through hydrolytic mechanisms.

3.4.2. Catalyst

The use of zeolite as a catalyst to assist in sludge decomposition was found to increase the production of gas but not improve oil and char yields due to the apparent increased conversion of volatile solid to gas. Sodium carbonate was also used as a catalyst to assist in sludge decomposition and the oil yield increased with the augment of catalyst.

Zinc chloride (Lu, 1995, 1996) can be used to degrade cellulose materials upon heating and subsequently causes dehydration with creation of pore structure. Optimum activating temperature is found to be 650°C with zinc chloride with BET surface area (309 m^2/g) and micro-pore area (171 m^2/g). The highest BET surface area was obtained at 5 M zinc chloride while the highest micro-pore area was at 3 M zinc chloride. The results show clearly that the sludge derived adsorbent is capable of removing H_2S by adsorption, which capacity is about 25% that of the commercial activated carbon.

3.5. SLUDGE PYROLYSIS KINETICS

3.5.1. Overview

Sludge is highly heterogeneous and complex and often has different compositions both qualitatively and quantitatively when it is obtained from different sources. Consequently, the energy potential and product properties of pyrolysis can vary widely. As a result, many studies have been conducted in an attempt to gain a better understanding of the thermal decomposition kinetics of sewage sludge. Dumpelmann (1991) used TGA to study the pyrolysis kinetics of sewage sludge and found that pyrolysis occurred as a result of a competitive reaction for volatile substances and intermediate solids, which subsequently decomposed into volatile substances and char. However, it has been difficult to use simple global kinetics model to evaluate sludge pyrolysis reactions over a wide temperature range. The thermal degradation of waste sludge occurs via a series of complex chemical reactions, which poses problems when evaluating the actual mechanism of pyrolysis. Indeed, the presence of overlapping reaction sequences prevents the use of a simple model designed by constructing the best straight line to describe the reaction being (Wilburn, 2000). Therefore, the sum of reactions and fractions of the sludge play an important role in determining the kinetics parameters.

Conesa (1997) proposed a scheme in which three independent fractions of sludge corresponding to biodegradable matter, dead bacteria and non-biodegradable matter were considered. Alternatively, Yong (2002) divided the pyrolysis process into high, medium and low temperature regions based on an assumption of wet sludge being comprised of water, two major degradable compounds and char. He found that the shrinking core model could be applied to analyze the char combustion reaction.

3.5.2. Thermal Decomposition Behavior of Sewage Sludge

Oil refinery waste activated sludge (sludge OR) was obtained from an oil refinery wastewater treatment plant that treats wastewater from a crude oil refinery industry by the traditional aeration process. The components of sewage sludge are shown in Table 3-2.

Table 3.2. the components analysis of petrochemical sludge

Moisture, %	Ash, %	Volatile matter, %	Fixed carbon, %	Low heating value (kJ·kg^{-1})
6.98	27.34	57.47	8.21	15260

Sludge OR contains high concentrations of organic substances and high levels of moisture. Next, the dried samples were crushed into fine powders and then analyzed using a thermal analyzer to create thermographs under a nitrogen flow of 60 ml/min. The samples were then heated at rates of 5, 10, 15 °C/min, respectively. The initial sample weight was controlled approximately at 5-8 mg for each run and the weight loss data were observed at temperature range of ambient to 900 °C, TG and DTG curves are given in Figure 3-4 and Figure 3-5.

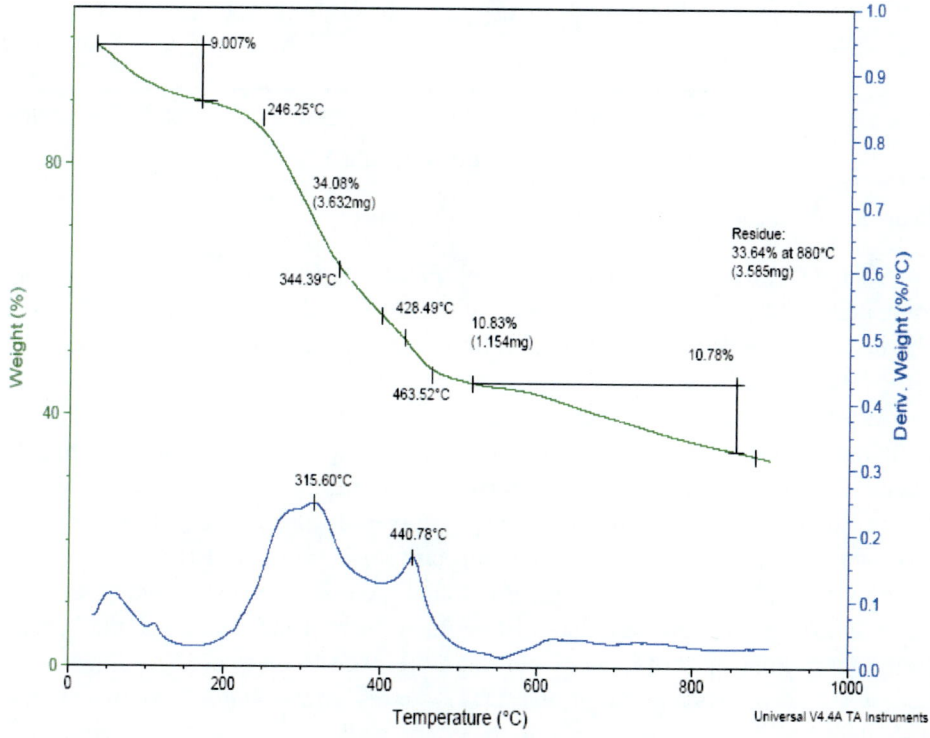

Figure 3.4. TG and DTG curves of sludge OR at heating 5°C/min.

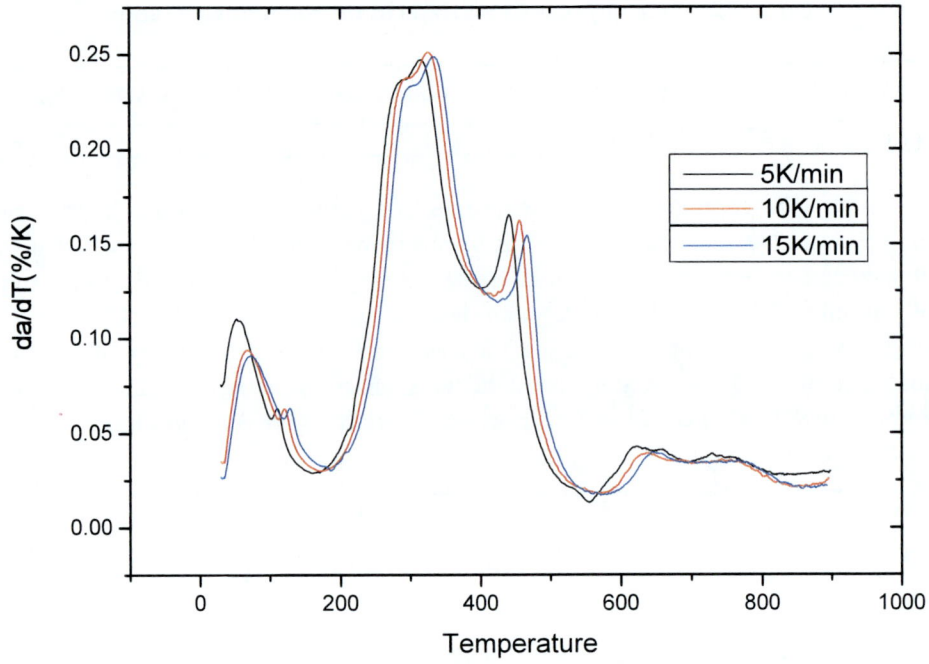

Figure 3.5. DTG curves of sludge RO at different heating rates.

Sludge OR lost most of its mass between 200 °C and 550 °C. The sludge OR contained a higher level of volatile substances and a lower level of salts. Therefore, the peak associated with volatile substances in the range of 200 °C - 550 °C gave prominence to the others.

As shown in Figure 3-5, the DTG curves gradually moved to the right, but maintained almost the same shapes as the heating rates increased. A higher heating rate should move the curves further to the right due to the material taking longer to reach that temperature at a lower heating rate. However, the decomposition curves exhibited anomalous crossings, particularly between 500 °C and 650 °C. Fisher et al. (2002) also found that the reactions above 400 °C exhibit many unique characteristics that differentiate them from reactions that occur during the primary pyrolysis step and therefore need to be treated separately. To gain a better understanding of the DTG curves corresponding to the sludge evaluated here, kinetics studies should be conducted.

3.5.3. Pyrolysis Kinetics Model

1. *Traditional Pyrolysis Model*
The equation of sludge pyrolysis can be written as.

$$\frac{dc}{dt} = k(T) \cdot f(c) \tag{1}$$

where t, T, and c are time, temperature and the concentration of the individual component i, respectively, and $f(c)$ represents the pyrolysis mechanism of component i. In addition, $k(T)$ can be given by the Arrhenius equation, $k(T) = Ae^{-\frac{Ea}{RT}}$, where E_a is the activation energy, A is the pre-exponential factor and R is the universal gas constant. When conducting solid-state kinetics analysis, α replaces c to express the extent of the reaction, which is defined as:

$$\alpha = \frac{W_t - W_o}{W_\infty - W_o} \tag{2}$$

where W_o, W_t, W_∞, are the weights or weight percentages of a sample during the initial, a certain time t, and final reaction, respectively. Temperature usually increases according to a certain constant, which is usually signified by the heating rate, β, which is given by $\beta = \frac{dT}{dt}$. Assuming that the pyrolysis reactions comply with nth-order reaction, Eq. [1] can be rewritten as :

$$\frac{d\alpha}{dT} = \frac{A}{\beta} \cdot e^{-\frac{Ea}{RT}} (1-\alpha)^n \tag{3}$$

α and $\frac{d\alpha}{dT}$ can be obtained from the TG and DTG curves respectively. Taking logarithms for component i, Eq. [3] can be expressed as:

$$\log\left(\beta \frac{d\alpha/dt}{(1-\alpha)^n}\right) = \log A - \frac{Ea}{2.303R} \cdot \frac{1}{T} \tag{4}$$

in which $\frac{d\alpha}{dT}$ is obtained from DTG curves at a certain value of α. A plot of $\log\left(\beta \frac{d\alpha/dt}{(1-\alpha)^n}\right)$ against $\frac{1}{T}$ should produce a straight line with a slope of $-\frac{Ea}{2.303R}$ and an intercept $\log A$, which can calculate the activation energy and pre-exponential factor and the optimal n.

Traditional pyrolysis model has been reported to use on sewage sludge pyrolysis kinetics analysis, but the values of the kinetics parameters were obtained based on the assumptions that the pyrolysis reactions comply with an nth-order reaction. Indeed, the kinetics parameters are not independent, but can affect each other through what are known as compensation effects. As a result, the values of kinetics parameters could change with changes in any of the assumptions.

2. Lorentz Fitting-multi-heating Method Model

A multi-heating rate method that was capable of deducing the actual reaction mechanism without any assumptions was chosen to interpret the sewage pyrolysis. The results presented could lead to a better understanding of the thermal decomposition kinetics of sludge. In the following, the Lorentz fitting method was introduced to separate the complex overlapping DTG peaks into several individual peaks to enable clarification of sewage sludge pyrolysis data in terms of the compositions of sludge and various gas evolutions produced during the sludge pyrolysis.

Assuming only one decomposition reaction occurs for a certain substance and there are no interactive reactions and function $f(c)$ is independent of temperature within a certain range of temperature. The kinetics equations for sludge pyrolysis can be written as a sum of the individual components' behavior.

$$\sum_i \frac{dc_i}{dt} = \sum_i k_i(T) \cdot f(c_i) \tag{1}$$

α_i replaces c_i to express the extent of the reaction:

$$\sum_i \alpha_i = \sum_i \frac{W_t^i - W_0^i}{W_\infty^i - W_0^i} \tag{2}$$

Eq. [1] can be rewritten as :

$$\sum_i \frac{d\alpha_i}{dT} = \sum_i \frac{A_i}{\beta} \cdot e^{-\frac{Ea_i}{RT}} f(\alpha_i) \tag{3}$$

α_i and $\frac{d\alpha_i}{dT}$ can be from the TG and DTG curves respectively. Taking logarithms for component i, Eq. [3] can be expressed as:

$$\log\left(\beta \frac{d\alpha_i}{dT}\right) = \log[A_i f(\alpha_i)] - \frac{Ea_i}{2.303R} \cdot \frac{1}{T} \tag{4}$$

in which $\frac{d\alpha_i}{dT}$ is obtained from DTG curves at different heating rates at a fixed value of α_i. A plot of $\log\left(\beta \frac{d\alpha_i}{dT}\right)$ against $\frac{1}{T}$ should produce a straight line with a slope of $-\frac{Ea_i}{2.303R}$, which enables the activation energy Ea_i to be calculated regardless of the function of $f(\alpha_i)$.

Eq. [4] can be rewritten as:

$$\log\left(\beta \frac{d\alpha}{dT}\right) + \frac{Ea}{2.303R} \cdot \frac{1}{T} = \log[A] + \log[f(\alpha)] \tag{5}$$

To determine the most probable mechanism, $f(\alpha)$, an optimal linear plot of $\log\left(\beta \frac{d\alpha}{dT}\right) + \frac{Ea}{2.303R} \cdot \frac{1}{T}$ against $\log[f(\alpha)]$ can be used to determine the reaction mechanism function $f(\alpha)$ and the pre-exponential factor A in Eq. [5]

The DTG peaks appeared to be overlapping and asymmetrical, which implied that several substances in the sludge may decompose together at nearly the same temperature range. Indeed, a better fit can be gained when there are a greater number of reactions. However, based on the compositions and the DTG curves of the sludge evaluated, it is likely that at least 5 pyrolysis reactions occurred during pyrolysis, which correspond to pyrolysis reactions of volatile substances, microbial cells, proteins, inorganic substances and fixed carbon. Moreover, the sludge OR at least 5 decomposition peaks were required to give an $R^2 \geq 0.99$ (Figure 3-6).

The α_∞ value, which indicated the percentage of a certain composition in sludge, was calculated based on the integration of each peak. There were no reactions capable of destroying the primary macromolecular constituents of the microbial cells at temperatures lower than 300°C and only the lighter compounds melted and volatilize at these temperatures. Therefore, peak 1 most likely occurred due to the decomposition of volatile substances. Sludge is a heterogeneous material with a random distribution of volatile materials. A reaction mechanism of random nucleus-formation and subsequent growth is acceptable in this case.

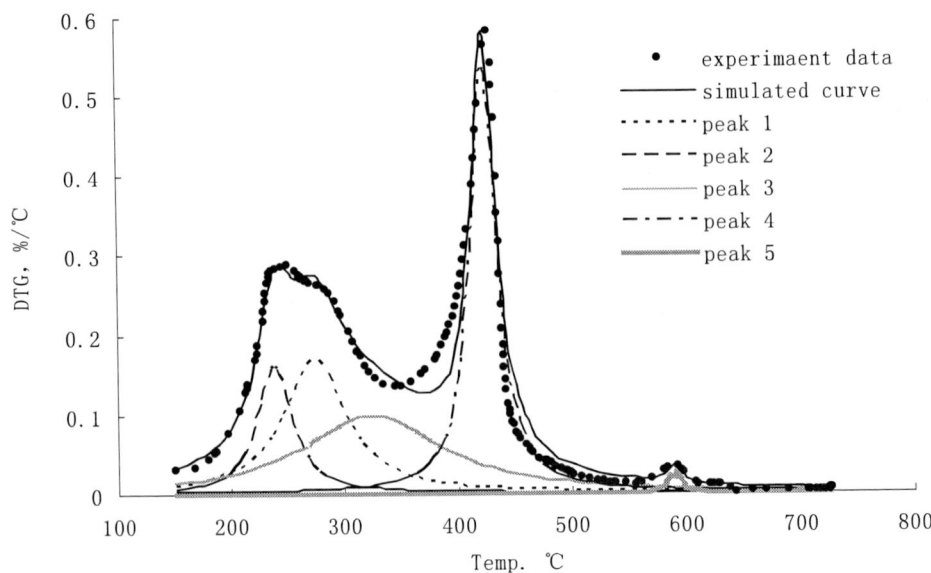

Figure 3.6. DTG decomposed curves of sludge at a heating rate of 5 °C/min.

The occurrence of peak 2 between 200 °C-450 °C may have been due to the decomposition of microbial cells. After volatile substances decompose, microbial cells begin to decompose by expanding in all three dimensions. At temperatures greater than 400 °C, the reaction is more destructive and macromolecules that contain peptide bonds, polypeptidoglycan and proteins begin decomposing (Chingguan, 2002). The highest concentration of total nitrogen was observed when the evolved gases were evaluated; therefore, peak 3 could represent the pyrolysis of those substances. Polypeptidoglycan and proteins are released first from microbes, at which point they undergo pyrolysis. As a result, the extent of the pyrolysis of the microbial cells likely influences the pyrolysis of polypeptidoglycan and proteins. Peak 4 was produced between 400°C -500°C, which indicates that the peak represents the decomposition of inorganic substances and the reaction of peak 4 showed a three-dimensional reaction mechanism. The high concentration of volatile substances resulted in the production of char during their pyrolysis period and sludge originally contained 8.21 % fixed carbon. Therefore, sludge showed a weak peak at approximately 590°C, signified as peak 5, which followed a third-order reaction mechanism.

REFERENCES

A. Ros, M.A. Lillo-Rodenas, E. Fuente, M.A. Montes-Moran, M.J. Martin, A. Linares-Solano, 2006. High surface area materials prepared from sewage sludge-based precursors. *Chemosphere* 65, 132–140.
Caballero J. A., Font R., Marcilla A., Conesa J.A., 1997. Characterization of sewage sludges by primary and secondary pyrolysis. *J. Anal. Appl. Pyrol.* 40–41, 433-450.
Charothon J., Vissanu M., Boonyarach K., Thirasak R., Pramoch R, 2007. Surface characterization and dye adsorptive capacities of char obtained from pyrolysis/gasification of sewage sludge, *Chemical Engineering Journal*, 133(1-3), 239-246.
Chingguan C., Hunglung C., Chihyu C., 2002. Pyrolytic kinetics of sludge from a petrochemical factory wastewater treatment plant—a transition state theory approach. *Chemosphere* 49, 431–437.
Conesa J. A., Marcilla A., Prats D., Rodriguez-Pastor M., 1997. Kinetic study of the pyrolysis of sewage sludge. *Waste Manage. Res.* 15:293-305.

Dumpelmann R, Richarz W, Stammbach MR., 1991. Kinetic studies of the pyrolysis of sewage sludge by TGA and comparison with fluidized beds. *Can. J. Chem. Eng.* 69,953–63.

Eun-Seuk Park, Bo-Sung Kang, Joo-Sik Kim. 2008. Recovery of Oils With High Caloric Value and Low Contaminant Content By Pyrolysis of Digested and Dried Sewage Sludge Containing Polymer Flocculants. *Energy Fuels* 22 (2), 1335–1340.

Fisher T., Hajaligol M., Waymack B., Kellogg D., 2002. Pyrolysis behavior and kinetics of biomass derived materials. *J. Anal. Appl. Pyrol.* 62, 331–349.

G. Q. Lu, D. D. Lau. 1996. Characterization of sewage sludge-derived adsorbents for H_2S removal. Part 2: Surface and pore structural evolution in chemical activation. *Gas Separation and Purification* 10(2), 103-111

G. Q. previous Lu, J. C. F. Low, C. Y. Liu, A. C. Lua, 1995, Surface area development of sewage sludge during pyrolysis. *Fuel* 74(3), 344-348.

Lilly Shen, Dong-Ke Zhang. 2003. An experimental study of oil recovery from sewage sludge by low-temperature pyrolysis in a fluidised-bed. *Fuel* 82(4) 465-472.

P. Thipkhunthod, V. Meeyoo, P. Rangsunvigit, B. Kitiyanan, K. Siemanond,T. Rirksomboon, 2005.Predicting the heating value of sewage sludges in Thailand from proximate and ultimate analyses. *Fuel* 84, 849-857.

P. Thipkhunthod, V. Meeyoo, P. Rangsunvigit, B. Kitiyanan, K. Siemanond, T. Rirksomboon, 2006. Pyrolytic characteristics of sewage sludge. *Chemosphere* 64, 955-962.

P. Thipkhunthod, V. Meeyoo, P. Rangsunvigit, T. Rirksomboon, 2007. Describing sewage sludge pyrolysis kinetics by a combination of biomass fractions decomposition. *J. Anal. Appl. Pyrol.* 79,78–85.

Wilburn F.W., 2000. Kinetics of overlapping reactions. Thermochim. Acta 354, 99-105

Yong Ho Yu, Sang Done Kim, Jong Min Lee, Keun Hoo Lee,2002. Kinetics studies of dehydration, pyrolysis and combustion of paper sludge. *Energy* 27,457–469.

INDEX

A

absorption, 5
acceptor, 13
acetate, 17
acetic acid, ix, 17, 22, 23, 24, 25
acid, ix, 5, 9, 15, 17, 23, 24, 25, 37
acidity, 35
activated carbon, 34, 35, 37
activation, 12, 18, 20, 22, 24, 34, 41, 42, 43, 46
activation energy, 18, 20, 22, 24, 41, 42, 43
adjustment, 5
adsorption, 13, 34, 35, 37
agents, 5, 10
agricultural, 3, 7
agriculture, 2, 3, 4, 6, 7, 28
air, ix, 9, 15, 16, 18, 20, 22, 23, 26, 27, 28, 29, 31
air pollution, 31
alcohol, 23
alkaline, 4, 5, 9, 14, 15
alkaline hydrolysis, 15
alternative, 1, 16, 31
alternatives, 10, 15, 16
amino, 17
amino acid, 17
ammonium, 25
anabolism, 6
anomalous, 40

application, 1, 2, 3, 4, 5, 6, 8
Arrhenius equation, 17, 21, 22, 41
assumptions, 42
atmosphere, 35
atmospheric pressure, 31
atoms, 23
availability, 4, 5, 8

B

bacteria, 5, 13, 38
behavior, 12, 42, 46
Beijing, 3, 4, 5, 7, 11
benefits, 3, 31, 34
binding, 12
bioenergetics, 6
biogas, ix, 7, 9, 10, 11, 15, 16, 24, 25, 26
biomass, 32, 33, 46
biopolymers, ix, 9
bonds, 32, 45
by-products, 16

C

cadmium, 12
calcium, 3, 34
carbamide, 3
carbohydrate, 10
carbon, 23, 26, 27, 34, 35, 36, 37, 39, 44, 45
carbon atoms, 23

carbon dioxide, 26, 27, 37
carbon monoxide, 37
catabolism, 6
catalyst, 37
cation, 4
cavitation, 14
cell, ix, 9, 10, 13, 14, 15, 16, 17, 27
cellulose, 34, 37
ceramic, 14
CH4, 10, 11
char combustion, 38
charring, 35
chemical agents, 5
chemical content, 7
chemical reactions, 38
chemicals, 15, 31
China, 1, 2, 3, 4, 5, 6, 7, 8, 10, 11, 29
chloride, 3, 34, 37
chlorine, 34
chromium, 12
Co, 13
CO2, 11, 17, 35
coal, 33
combustion, 31, 38, 46
compensation, 42
competition, 13
complexity, 13
components, 6, 13, 23, 37, 38, 39, 42
composition, 10, 12, 13, 33, 44
compost, 6, 8, 26
compounds, 10, 15, 17, 20, 32, 38, 44
concentration, 3, 4, 8, 10, 12, 13, 14, 18, 21, 23, 33, 35, 36, 37, 41, 45
configuration, 28
construction, 31
consumption, 5, 16
contaminant, 2, 34
contaminants, 2, 12
contamination, 3, 5
control, 2, 4, 5, 8, 24, 31
conversion, 31, 32, 37
conversion reaction, 31
copper, 12
correlation, 20
costs, 3, 24, 31

cracking, 33
crops, 4
crude oil, 31, 34, 38
cyclone, 34

D

decomposition, 6, 31, 32, 37, 38, 40, 42, 44, 45, 46
decomposition reactions, 32
degradation, 9, 14, 28, 38
degradation process, 9
dehydration, 35, 37, 46
denaturation, 14
denitrification, 12
density, 26
destruction, 16
detergents, 12
developed countries, 3, 10
developing countries, 10
diesel, 32
diesel fuel, 32
diffusion, 17
digestion, ix, 3, 9, 10, 11, 13, 14, 15, 24, 25, 26, 27, 28, 29, 34
distillates, 32
distribution, 6, 44
dosage, 2, 4, 5, 15
dumping, 2

E

economic development, 6
effluent, 25, 26
electron, 13
electrostatic interactions, 35
energy, ix, 1, 4, 6, 9, 14, 20, 22, 24, 25, 31, 32, 38
engines, 32
environment, 3, 5
environmental protection, 6
enzymes, 12, 13
equilibrium, 26, 27, 35
Ethanol, 23

Europe, 1
European Union, 2, 3
eutrophication, 4
evolution, 33, 46

F

farmlands, 4
fat, 10
fatty acid, 17, 25
fatty acids, 25
fermentation, 12
fertilizer, 3, 5
fertilizers, 3, 4
fiber, 10
fillers, 34
filtration, 25
fixation, 28
flow, 39
fluidized bed, 34, 46
food, 10
Forestry, 7
fossil, ix
fossil fuel, ix
free radical, 26
freezing, 9, 26
FT-IR, 35
fuel, ix, 34

G

gas, 15, 17, 18, 20, 32, 33, 37, 41, 42
gas phase, 17
gases, 45
gasification, 45
generation, 16
glass, 14
government, 1
groups, 13, 20
growth, 44
Guangzhou, 8, 11

H

H_2, 11
harm, 2, 4
heat, 15, 16, 24, 25, 32, 33
heat capacity, 24
heating, 15, 24, 31, 32, 33, 34, 35, 37, 39, 40, 41, 42, 43, 44, 46
heating rate, 32, 33, 40, 41, 42, 43, 44
heavy metal, 2, 3, 4, 5, 6, 8, 12, 13, 14, 31
heavy metals, 2, 3, 4, 5, 8, 12, 13, 14, 31
heterogeneous, 10, 38, 44
high pressure, 28
high temperature, 9, 15
hydro, 31, 32
hydrocarbons, 31, 32
hydrogen, 14, 34
hydrolysis, ix, 9, 14, 15
hydroxide, 14
hydroxyl, 16

I

incineration, 1, 31
industrial, ix, 1, 10, 12, 13, 24
industry, 38
inhibition, 12, 13, 26
inhibitory, 6, 12
injury, vi
integration, 44
interactions, 35
IR spectra, 35
irradiation, 16, 27

K

kinetic energy, 14
kinetic model, ix, 17, 18, 19, 20, 21, 22, 23, 24, 28
kinetic parameters, 22
kinetics, ix, 26, 27, 38, 40, 41, 42, 45, 46

L

land, 1, 2, 3, 4, 5
landfill, 1, 3
landfills, 3
Langmuir, 35
large-scale, 14
law, 2
laws, 2, 4
leaching, 5, 12
legislation, 1
limitation, 17
linear, 43
lipids, 31
liquid phase, 17
liquids, 32
low-temperature, 32, 46

M

macromolecules, 45
magnesium, 3
management, ix, 1, 2, 3, 4, 6
manure, 7
meat, 10
metabolism, 14
metal content, 2, 11, 12
metals, 2, 3, 4, 5, 8, 12, 13, 14, 25, 31, 34
methane, 10, 15, 24, 26
methanogenesis, 13, 14
microbes, 45
microbial cells, 14, 44, 45
microorganisms, 6, 12, 14
milk, 10
mineralization, 4
models, 6, 27
moisture, 39
mole, 13
molecular weight, 10, 15
molecules, 13, 16, 35
monolayer, 35
movement, 5
municipal sewage, 8

N

NCG, 32, 33
neutralization, 15
Ni, 5, 8, 11, 13, 25
nickel, 12
nitrate, 5, 13
nitrification, 12
nitrogen, 3, 4, 10, 12, 25, 34, 37, 39, 45
N-N, 12
normal, 3
nucleus, 44
nutrient, 2, 3, 4, 5, 10, 11

O

oil, 19, 31, 32, 33, 34, 37, 38, 46
oil production, 37
oil recovery, 32, 46
oils, 34
organic compounds, 17, 20
organic matter, ix, 3, 9, 15, 37
oxidation, ix, 9, 10, 15, 17, 20, 23, 25, 26, 27, 28, 29
oxidation products, 20
oxide, 34, 36, 37
oxygen, 13, 17, 18, 21, 25, 31, 32
ozonation, 9, 15
ozone, 16

P

parameter, ix
pathogens, 2, 4, 9, 16
Pb, 5, 8, 11, 12, 13
peptide, 45
peptide bonds, 45
peroxide, 16
petrochemical, 20, 22, 23, 39, 45
petroleum, 32
petroleum distillates, 32
pH, 4, 5, 9, 14, 15, 16
phenol, 26, 27, 28

phenol oxidation, 28
phosphate, 3
phosphorus, 3, 5, 10, 25
physical properties, 26
plants, 3, 5, 7, 8, 10, 12, 33
play, 38
pollutant, 2
pollution, 2, 4, 5, 7, 8, 28, 31
polymer, 34
poor, 9
pore, 34, 35, 37, 46
potassium, 3, 10, 11
powders, 12, 39
power, 9, 15, 16
precipitation, 6, 13, 14
pressure, 14, 28, 31
prevention, 4
producers, 26
production, ix, 1, 6, 7, 9, 11, 14, 15, 16, 24, 25, 26, 31, 32, 34, 37, 45
property, vi, 10
protection, 1, 6
protein, 10, 13, 14
proteins, 14, 31, 44, 45
pyrolysis, 25, 31, 32, 33, 34, 35, 36, 37, 38, 40, 41, 42, 44, 45, 46
pyrolysis reaction, 32, 38, 41, 42, 44

R

radical mechanism, 26
random, 44
range, 14, 32, 34, 37, 38, 39, 40, 42, 44
reaction mechanism, 42, 43, 44, 45
reaction rate, 17, 19, 20
reaction temperature, 17, 21
reclamation, ix
recovery, 27, 31, 32, 46
refineries, 32
refractory, 20
regulation, 2
relationship, 6
relationships, 12
relative toxicity, 13
residues, 25, 31

resistance, 17
retention, 9, 25, 32
rice, 3, 4, 5

S

salts, 40
sample, 32, 34, 39, 41
sampling, ix
sand, 3
SBR, 14
secondary inhibition, 13
selectivity, ix, 17, 21, 22, 24
sensitivity, 13
separation, 34
series, ix, 6, 38
services, vi
sewage, ix, 1, 3, 6, 7, 8, 9, 10, 12, 13, 15, 16, 26, 28, 29, 31, 32, 33, 34, 35, 38, 42, 45, 46
Shanghai, 11
shear, 14
sites, 3
skeleton, 35
sludge, ix, 1, 2, 3, 4, 5, 6, 7, 8, 9, 10, 11, 12, 13, 14, 15, 16, 17, 18, 19, 20, 22, 23, 24, 25, 26, 27, 28, 29, 31, 32, 33, 34, 35, 36, 37, 38, 39, 40, 41, 42, 44, 45, 46
sodium, 17
soil, 2, 3, 4, 5, 6, 8
solid waste, 1
solid-state, 41
solubility, 15
species, 13
specific surface, 33
speed, 6, 14, 19
stabilization, 28, 31
stabilize, 9
stages, 14
standards, 2
starch, 10
steel, 14
storage, 32
strains, 13
strategies, 14

substances, 2, 3, 10, 15, 17, 20, 22, 38, 39, 40, 44, 45
substrates, 13, 31
sugar, 10
sugarcane, 3
sulfate, 13, 26, 27, 28
sulphur, 13, 25
supernatant, 23
supply, 7
surface area, 34, 35, 37, 45
surface water, 5

T

temperature, 9, 18, 20, 22, 24, 25, 32, 33, 34, 35, 37, 38, 39, 40, 41, 42, 44, 46
temperature dependence, 22
TGA, 38, 46
thawing, 9, 26
thermal decomposition, 31, 38, 42
thermal degradation, 38
thermal treatment, 9, 15, 16, 26
thermolysis, 31
TOC, ix, 17
toxic effect, 12
toxicity, 13
transfer, 14, 17
transition, 45
transport, 17
transportation, 2
turbulence, 14

U

ultrasound, 27
universal gas constant, 41

universities, 4
urbanized, 6

V

values, ix, 22, 24, 34, 42
vapor, 33
vegetables, 8
viscosity, 34
volatile substances, 38, 40, 44, 45
volatilization, 12

W

wastewater, 1, 3, 4, 6, 7, 10, 12, 13, 20, 26, 29, 33, 38, 45
wastewater treatment, 1, 3, 4, 6, 7, 10, 12, 26, 33, 38, 45
wastewaters, 13, 17, 28
water, 2, 4, 5, 7, 16, 17, 24, 28, 29, 38
water vapor, 24
weakness, 2
weight loss, 34, 39
wheat, 5

Y

yield, 3, 10, 26, 33, 37

Z

Zinc (Zn), 5, 11, 12, 13, 25, 34, 35, 37